MECHANICAL RELIABILITY AND DESIGN

MECHANICAL RELIABILITY AND DESIGN

A.D.S. Carter

BSc, CEng, FIMechE, FRAeS
Professor Emeritus,
Royal Military College of Science
Hon Fellow, University of Wales, Swansea

First published 1997 by
MACMILLAN PRESS LTD
Houndmills, Basingstoke, Hampshire RG21 6XS
and London
Companies and representatives
throughout the world

ISBN 0–333–69465–1

A catalogue record for this book is available
from the British Library.

This book is printed on paper suitable for recycling and
made from fully managed and sustained forest sources.

10   9   8   7   6   5   4   3   2   1
06   05   04   03   02   01   00   99   98   97

Printed in Hong Kong

Contents

Preface

Following the publication of my earlier book in 1986 I lectured to a number of specialist courses, seminars and the like on various parts of its contents. Gradually, these lectures concentrated more and more on the design aspects of mechanical reliability, which many saw as the key area. Having retired from this activity, I have brought together the final form of these various lectures in a more developed format in this book. The background has been filled in more fully than was possible in the limited time for lectures, and continuity material has been added. The opportunity has been taken to dot a lot of i's and cross many t's of the supporting theory so as to produce a coherent picture of the role of reliability, particularly as a quantitative requirement, in the design of mechanical machines and components and, in particular, their constituent parts.

Like the lectures on which it is based, the book is aimed at practising mechanical engineers; but, like the lectures, it could be used in specialist courses and seminars. Additionally, parts of it could well be included in courses on mechanical design at first degree level. I do not see how the education of any mechanical engineer can be considered complete without some reference to quantified statistical reliability, and to the impact of specifications for such reliability on design.

The book makes full use of statistical reliability concepts, but crucial fundamental aspects are critically reviewed before use. In the event it was found that surprisingly little statistics is actually needed for a design methodology aimed to meet a quantified reliability, but it was found necessary to delve a little further into materials science than is customary. I hope that the reader, having read the book, will accept that a design methodology based on statistically defined input data to meet a quantified reliability specification is feasible, not too different from contemporary practice, and not difficult to implement. I hope too that he or she will be prepared eventually to say goodbye to all the uncertain empiricism that surrounds so much of contemporary design.

Finally, I wish to acknowledge my indebtedness to all those reliability workers who have carried out and reported stochastic test data on materials and components, both those quoted in this book and those consulted but not quoted.

Full reference has been given in the text to all quoted data, so that the reader may consult the original work. Obtaining worthwhile data is very time consuming and laborious, but those data are the bedrock of all our knowledge. My thanks to them all, especially to British Coal for permission to use previously unpublished data in the case study of Chapter 7.

A.D.S. Carter
Oxford 1996

Notation

A	Material constant (microstructurally short crack)
A_c	Empirical constant in creep rate equation
a	Crack length
B	Material constant (physically small crack)
C	Physically small crack threshold
c	Cost
c_i	Constants
D	Material constant (physically long crack)
d	Distance to barrier
d	Damage (Section 6.1)
E	Damage resistance threshold
$E_N(s)$	PDF of damage resistance at N load applications
$E(s)$	PDF of damage resistance threshold ($N \to \infty$)
$\bar{E}$	Mean value of $E(s)$
F	Cumulative probability of failure
$F(x)$	CDF of failure
$f(x)$	PDF of failure
G	Miner summation
g	Number of entanglements
$H(t)$	Cumulative hazard function
h	Hazard
$h(t)$	Hazard expressed in terms of t
$h(n)$	Hazard expressed in terms of n
I	Number of intervals
i	$1,2,3,\ldots$, etc.
j	$1,2,3,\ldots$, etc.
K	Scaling factor
k	Number of standard deviations from mean to worst case
L	Load
$L(s)$	PDF of failure-inducing stress due to load
$\bar{L}$	Mean value of $L(s)$
$L(p)$	PDF of load

$\bar{L}(p)$	Mean value of $L(p)$
LR	Loading roughness
l	Number of loads
M	Maintainability
MP	Mean probability (mean rank)
Mz	A Weibull function
m	Index in power relationship
N	Number of load applications to failure
n	Number of load applications
P	Probability
p	Force or other impressed phenomenon
Q	Total number of items in population
Q_c	Activation energy for creep
q	Number of items
R	Reliability (cumulative probability of no failure)
R	Universal gas constant (section 6.2.2)
$R(n)$	Reliability expressed in terms of n
$R(t)$	Reliability expressed in terms of t
$R(c)$	Rockwell – c scale
S	Strength
$S(s)$	PDF of strength at critical section
$\bar{S}$	Mean value of $S(s)$
SM	Safety margin
s	Stress
T	Time to failure
t	Time
Σt	Cumulative operating time
t'	Time from different origin
t_o	Weibull locating constant
U	Ultimate tensile strength
$\bar{U}$	Mean ultimate tensile strength
u	Cross-sectional area
$V(s)$	PDF of strength modification during manufacture
$\bar{V}$	Mean of $V(s)$
W	Number of breakdowns
ΣW	Cumulative breakdowns
X	General arbitrary material property
x	General arbitrary variable/variate
Y	Yield strength
Z	Attribute
z	$s_F - s_{50}$
α	Material constant (microstructurally short crack)
β	Material constant (physically small crack)
β	Weibull shaping parameter

γ	Coefficient of variation
ε	Strain
$\Delta\gamma_p$	Shear plastic strain range
ΔK	Stress intensity factor
ζ	s–N function
η	Characteristic life
θ	Number of tests
λ	Breakdown rate
μ	Mean value
ν	Duane index
σ	Standard deviation
τ	Temperature
ϕ	Factor of safety
ω	Maintenance cost index

Subscripts

a	Average
c	Refers to strength collapse
c	Value at confidence limit
d	Defects
E	Refers to damage resistance threshold
F	Value for $F\%$ failures
g	Refers to entanglements
i	Successive values $1,2,3,\ldots$, etc.
j	Successive values $1,2,3,\ldots$, etc.
l	Lower limit
L	Refers to load
m	Of mean
max	Maximum perceived value
min	Minimum perceived value
N	Value at N load applications
nom	Nominal value
O	Reference or arbitrary value
o	Original or initial value
op	Operational
p	Value for process
S	Refers to strength
SH	Refers to strain hardening
U	Refers to ultimate tensile strength
u	Upper limit
Y	Refers to yield strength
0	Lowest acceptable value
*	Minimum physically small crack

Index

*   Specified value

Note: SI units are used throughout, except that original data have been quoted as such, i.e. in the originator's units. The SI equivalent is always given when the quantity could have general significance.

Glossary

assembly *See* **hierarchy of mechanical artefacts.**

break down To rupture the continuity of performing a given function. To disrupt or stop for a time.

breakdown rate The rate at which breakdowns occur.

component *See* **hierarchy of mechanical artefacts**.

damage A change in a material structure.

damage resistance The ability of a material to resist damage by a load or other hostile environment.

fail To become exhausted; to come to an end.

failure The termination of the ability to perform a required function. The inanimate equivalent of death.

failure rate Term not used in this book.

field 'In the field' or 'in service' implies that the product is in the hands of the intended user, and data so described refer solely to his/her/their use of the product. Tests or trials of any kind are specifically excluded.

hazard The non-dimensionalised rate at which failures occur in a non-maintained component or part. For the mathematical definition see section 3.5 (equation 3.36).

load Any hostile environment that can give rise to a failure. See Chapter 1 for a comprehensive definition.

hierarchy of mechanical artefacts There is no standardised terminology, but the following definitions apply to their usage in this book. (Given in hierarchical order, not alphabetical.)

    system A combination of integral machines and components to achieve a specified output(s).

subsystem   A convenient subset of a system made on various different criteria, such as providing one of several different inputs to the system, operating on an identifiable basis (e.g. hydraulic), contained as an integral package, provided by a given supplier, and so on, Can be extended to sub-subsystem, etc.

machine   An integral assembly of components and parts to achieve a defined output or related series of outputs. The lowest subsystem level, but could well be a stand-alone artefact.

assembly   A package of components and parts that comprises the machine. The package is usually determined by manufacturing or handling convenience rather than any functional attribute, though they may well coincide. Can be extended to sub-assembly, etc.

component   An identifiable self-contained unit, to perform a given function, made up of several or many parts. The lowest assembly level. It may or may not be maintainable.

part   The fundamental mechanical artefact, which cannot be dismantled. Essentially, it is completely defined by a detail working drawing – to use old nomenclature – or by a modern electronic record or hard copy. Hence often referred to as a 'detail' in the literature. The word 'piece' is also often used, especially for a part made specifically for test (testpiece). A part or piece is not maintainable except by replacement.

product   Can be any of the above, but generally excluding systems. The final output of a manufacturer or manufacturing organisation.

item   A single constituent of a statistical population on which an observation or measurement is made. Sometimes called 'specimen' in statistical literature.

machine   *See* **hierarchy of mechanical artefacts.**

operational reliability   The ability of a system/product to perform the required function under stated conditions of use and maintenance for a stated period of time.

part   *See* **hierarchy of mechanical artefacts**.

population   The totality of items possessing specified attributes. The whole product.

prototype   The first or primary version of any product. It is used to establish the feasibility of the product and as a model for subsequent full production. There may be several prototypes.

product   *See* **hierarchy of mechanical artefacts**.

reliability The ability of a system/product to perform the required function under stated conditions of use for a stated period of time.

ROCOF Acronym for the rate of occurrence of failures; equivalent of breakdown rate.

sample A subset of a population.

service *See* **field**.

standard *s–N* curve The *s–N* curve derived by contemporary conventional practices. It is truly neither the mean nor the median curve, but is closer to the latter.

strength The ability to resist a compatible load without failure at any given instant. A product may have several different strength attributes.

stress-rupture A failure mechanism in which failure, rupture, or fracture of the material may take place on application of a load, but if not the material structure remains unchanged.

system *See* **hierarchy of mechanical artefacts**.

threshold The limit state above which damage may be inflicted and below which no damage can be done.

variation A difference in a property of the items making up a population. It includes both the statistical variance (or standard deviation) and the shape of the distribution.

wear A failure mechanism in which no failure, rupture, or fracture necessarily takes place on the application of a load, but there is a change in the material structure (i.e. damage is done). Failure takes place when the cumulative damage reaches some critical value.

wear-out The period of time during which wear failures take place.

1
Introduction

This book is concerned with those aspects of the design of mechanical machines, and hence the components and parts from which they are made, that ensure reliability. But, because the words 'design' and 'reliability' can mean so many different things to different people, it is necessary to be quite clear at the outset exactly how those words will be used in this book. To start with, we shall exclude from our treatment all aspects of functional design. Thus it will be assumed that at least one prototype has demonstrated that it is possible to build a machine, or a component of a machine, that will do the job we require it to do to our satisfaction. We shall also exclude all aesthetic design, or styling, from the treatment, even though to many people this is what design is all about. This book will then be concerned with the design methodology that ensures that a number of items of the same product, i.e. made to the same design, will perform in the same way whenever operated in the same manner: that is, there shall be no, or in some case only an acceptable number of, adventitious malfunctions, breakdowns or failures. We shall thus be concerned with the inherent ability of the machine to continue functioning as specified. This may be thought of as a rough and ready definition of mechanical reliability.

Of course, reliability is an age-old concept, and everybody thinks he or she knows what they mean when they say that a thing is reliable. But that age-old concept is a bit vague when examined critically, and in the last quarter of a century or so attempts have been made to be more precise by basing definitions of reliability on statistical probabilities. This has stemmed from a new technological discipline, which might loosely be called the science of reliability, that has developed in the period. It is manifest in a vast literature that has grown up on the subject of reliability. An examination of this literature suggests that the new subject is far removed from the common man's idea of reliability: it has become a subject of intense mathematical study, closely associated with statistics. But it goes beyond that. No longer do sophisticated customers demand 'reliability'; they now quantify their requirements with statements along the lines that the machine they wish to purchase must be at least 99.9 per cent (or some other appropriate figure) reliable at a specified age or for a specified period of operation. How does the designer meet this new demand? Can existing design techniques be adapted to provide the necessary assurance? Or must new design techniques, based on the developing statistical approach, be adopted? This book is a search for the answers to such questions.

Such a search is necessary because a careful examination of the extensive literature on reliability shows that little has been written on the role of basic design in achieving reliability. I find this very surprising indeed. So far as mechanical reliability is concerned there do seem to me to be only two – possibly three – activities from the manufacturer's point of view that can possibly control the reliability of a product. The first is its design, i.e. the specification of its physical format, and the second is the quality control of its manufacture, which defines the extent to which the finished product conforms to that design. I should have thought this was self-evident: once a batch of items physically exist, there is no way in which their reliability can be changed, other than by remaking the product or modifying it. This leads to the possible third activity that can control reliability before the product leaves the manufacturer's hands. It would be unusual, though not unknown, to mass produce a product directly from an initial design. Prototypes would be made and tested, both for performance and (one hopes) for reliability. If the reliability were unsatisfactory, the deficiency should be rectified before the design was finalised. The modification process is usually referred to as 'development', but it necessarily involves a considerable amount of redesign, and thus is not a stand-alone activity. It is, of course, not an option for a 'one-off' product.

Development consists of operating prototypes in the same manner in which production items will be used in service, so far as that is possible, though it is usual to concentrate on the most adverse circumstances. Any inability to meet the design specification for performance or reliability is noted and rectified, or 'fixed' in the jargon, before the item goes into production. A fix would consist of an assessment of the reasons for the shortcoming, which may require some sophisticated techniques, and then a redesign to eliminate the problem. Design is an essential ingredient of development; indeed, development is more closely associated with design than with research, as is suggested by the universal acronym 'R&D'.

Leaving design for a moment, it is significant that the mathematics of statistical quality control, the second activity involved in achieving reliability, is highly developed. Organisations practising modern quality control make use of the statistical concepts that form part of the new discipline. This is as it should be, for if an acceptable design is not manufactured in accordance with the design specification, some loss of reliability must be expected. Taking it a stage further, if the design and/or quality control are inadequate, and an unreliable product does find its way on to the market, then that unreliability can be countered to some extent by maintenance. It is sometimes claimed that an unreliable design that can be produced and maintained cheaply could be a better economic proposition than a more reliable but more expensive design. This claim will be pursued more rigorously later. For the present, it is to be noted that some planned maintenance of mechanical products will always be required to counteract the effects of normal wear and tear, and should be provided for in design. But additional maintenance to remedy a reliability shortfall has to be paid for, not only in the actual cost of the additional maintenance itself, but also in the loss of output from the product during that maintenance process. Unplanned maintenance is not therefore a substitute for a design that does not achieve its specified reliability.

The optimisation of maintenance is well documented. The mathematical modelling, involving the statistics, of maintenance processes is apparently highly developed. Considerable theoretical effort has been devoted to this topic, though it is usually confined to an evaluation of an existing situation, i.e. one in which the product has already been designed, manufactured, and is in current use. To the extent that maintenance should be defined at the design stage, this is often a case of 'shutting the stable door after the horse has bolted'. A second-rate product is still second rate even when subject to optimum maintenance.

Is it not strange that the statistical theory of reliability is so highly developed for the two activities that succeed design, but almost non-existent, as revealed by any literature survey, for design itself, which precedes them and without whose successful implementation they are deployed in vain? Why should this be so?

The reason, I believe, lies in one of the fundamental assumptions underlying nearly all statistical reliability theory, and this assumption must be fully recognised when applying statistics to design. General conventional statistical reliability theory operates at what has been described as the epiphenomenological level (see for example Mallagh, 1988). By this is meant the absence of any reference in the theory to the causes of the events represented by the probabilistic distributions. In other words the modelling is a-causal (i.e. not causally related to any failure mechanism(s)). By contrast, in basic mechanical design we are essentially dealing with the phenomena themselves. The two methodologies differ in fundamental concept. An example may make this more clear.

Suppose we are designing a self-contained power supply for some application – an offshore rig for example. It may be thought of as a gas turbine prime mover driving an electrical generator plus the appropriate control gear. Then conventional reliability theory would state that the failure rate of the power supply system is the sum of the failure rates of the gas turbine, the generator, and the control gear (assuming them all to be constant). Or, the reliability at any time is the product of the reliabilities of the three subsystems at that time. If this does not meet the customer's requirements, we can consider duplicating the control gear (which can be expected to have the highest failure rate), i.e. we can employ a degree of parallel redundancy. If this is still inadequate, we can consider duplicating the whole set, or perhaps installing three half-power sets and so on – there are a number of possible alternatives.

In this kind of calculation, the failure rate for the gas turbine – on which I should now like to concentrate – would be specified at some expected value, typically at about one unscheduled removal per 10 000 running hours for a mature machine. Such data for all the subsystems enable the calculation outlined above to be carried out quite easily. Note that the figure of 1 per 10 000 hours enumerates the number of 'failures' without stating what those failures are, i.e. the data are a-causal (or epiphenomenological), and the appraisal of the system is carried out at that level. In the same way, if we have any system comprising a number of discrete subsystems that do not interact (i.e. are statistically independent) and for which failure data exist, then it is possible to design the system using various arrangements of those subsystems, choosing the one that meets the specification. This is the epiphenomenological methodology. Some authorities do not recognise the a-causal/epiphenomenological nature of their methodology, but do recognise that the methodology applies only to systems that can be constructed from 'blocks', which are statistically independent subsystems (see for example Bourne, 1988).

Whatever approach is adopted, however, the failures that have to be minimised are not in fact abstract events; they are real physical happenings of various kinds. Thus in the case of the gas turbine a 'failure' could be the failure of a compressor blade, a combustion chamber flame tube or injector, the fuel supply, a turbine blade; it could be a disc, though more likely a bearing or seal of some kind; or it could be one of innumerable other possibilities. All these failure mechanisms are mixed up higgledy-piggledy (randomly in statistical terminology) in the figure of 1 in 10 000 running hours. However, and it is a big 'however', to achieve reliability at the design stage it is necessary to deal with each of these failure modes at the physical level. Every one must be eliminated, using engineering not statistical techniques.

To pursue in greater depth the example taken: the design of the compressor blades of the gas turbine will model them as cantilevers (albeit of complicated shape to meet their aerodynamic function) having appropriately defined end-fixing and subject to a distributed (in the engineering not statistical sense) gas-bending load. It is also necessary to allow for any buffeting load to be expected from possible intake flow maldistribution, wakes from preceding rows, struts and so on. From such information the size of the blade can be calculated, using conventional causal stress analysis techniques. At the same time, it will be necessary to consider other failure mechanisms: for example, selection of the blade material to hold the blade profile in any specified corrosive/erosive atmosphere in which the unit has to operate (an essentially causal relationship). It will also be necessary to consider the vibration properties of the blade in relation to all exciting sources. The blade tip clearance must be chosen to avoid or accommodate any rub during the planned life. Then the blade root has to be designed, and so on. All this has to be done within cost constraints and the manufacturing resources available.

It is clear that all this is very much at the phenomenological level of modelling, involving causal relationships. Epiphenomenological modelling is simply ineffective and irrelevant to this activity. There is no way in which a figure of one failure in 10 000 running hours can be converted into a blade chord (keeping the same example) without applying causal physical relationships. This distinction between epiphenomenological and phenomenological methodology must be fully understood if statistical reliability theory is to be applied to the complete design process. Epiphenomenological methodology allows us to design systems from discrete subsystems (the 'building blocks') for which failure data exist, but phenomenological methodology is essential to design the individual bits and pieces that make up all subsystems. It should never be forgotten that no matter how complex and sophisticated our system may be, its reliability depends essentially on the reliability of all those bits and pieces. With this in mind, we can return to the problem of designing for reliability.

First of all, it is necessary to define exactly what is meant by 'reliability'. British Standards define it as 'the ability of an item to perform the required function under stated conditions for a stated period of time'. This is only one of a large number of definitions, which are all, however, very similar in nature. It is possible to quibble over the precise wording, but this is as good a starting point as any for our purpose. No units or numerical criteria are given for the attribute 'ability' in the BS definition. In statistical reliability theory and practice 'the ability to perform' is measured as the 'probability of performing', and is a number lying between 0 and 1 (or 0 per cent and 100 per cent). Reliability, then, assumes all the properties of probability as defined and developed in standard statistical theory. This is of course a great advantage, because the vast body of knowledge contained in that subject is immediately applicable to reliability. However, in whatever way 'ability' (i.e. reliability) is defined, it can be achieved only by making a mechanical part of a certain size and shape and from a certain material. Once these three quantities are specified, the part is completely defined, and so too is its reliability. Indeed, we make parts of specified sizes, shapes and materials for no other reason than that they should 'perform the required function under stated conditions for stated periods of time'.

To illustrate this with a simple example: suppose we have a shaft to transmit a specified power at a specified speed and that it is 40 mm in diameter when made in a defined material. The only reason why it is 40 mm in diameter is that it should be able to perform the specified function – otherwise there would be every reason for making it smaller. Contemporary design techniques do thus have reliability, measured subjectively rather than numerically, as an objective. Conversely, design for reliability is nothing more, nor less, than that process that the normal mechanical engineer understands by design. Design for reliability is not some mysterious add-on activity that has to be brought into our normal activities. To design for a quantified reliability we must fully assimilate quantification into conventional contemporary design practice.

Continuing with this illustration: given its material properties, the shaft dimension to transmit a given power at a given speed could easily be deduced by standard textbook methods. The difficulty is that in real life the power and speed will *not* be known – not known exactly, that is – and neither will the material properties be known exactly. These quantities are all statistically distributed, not deterministic. So, one shaft will not be identical to another of the same design, and may not be subject to the same load. If there were no variation at all, every item of a product would have the same life and be 100 per cent reliable up until the end of that life. With variation in these quantities, lives will also vary, so some items may fail before their planned life is achieved: i.e. the reliability is less than 100 per cent. Others may outlive their planned life. Variability is the source of unreliability.

Conventional design is based on the assumption that all quantities contributing to the design are deterministic and so can be specified uniquely. The effects of any and all variations are then catered for by empirical factors based on experience – often very long experience. In the statistical approach, variations in all the basic quantities contributing to the design are recognised at the outset, and are represented by established statistical parameters. Ideally, the parameters can be fed into the equally well-established design equations used to evaluate stress and strength, eliminating the need to call up any empirical factors. Because statistical techniques have been specifically developed to deal with quantities that are not deterministic but distributed in nature, this may be supposed to be a superior approach. Unfortunately, experienced designers will not use it. One reason is, I believe, that not being brought up in the use of mathematical statistics – which can become complicated if all the quantities involved in design vary independently – they have no 'feel' for the statistical quantities involved. (Equally I find that statisticians have no 'feel' for the physical quantities involved.) Design is a subjective iterative process, which is difficult enough without introducing further unknown (to designers) quantities. I also find that many designers have a 'gut feeling' that there is something not quite right with statistics – the 'lies, damn lies, and statistics' school of thought – perhaps not without some justification!

Turning to more tangible aspects, design is much more complicated than most textbook solutions suggest. It is not difficult to evaluate the torque in the shaft quoted above. Most engineers would realise that reversed bending may also be involved, and that it can easily be evaluated, so that the maximum combined stress and hence the shaft diameter can be calculated. But this is hardly likely to be enough. Other factors may be involved: for example, fitments will most certainly have to be accommodated. The shaft is no longer a cylindrical piece of metal, and the stress pattern within the shaft is changed. Stress raisers rear their ugly head. Although the broad stress distribution can be evaluated by modern stress analysis techniques, the stress raisers are themselves a source of marked and feared variation – a variation that is usually ignored by modern stress analysis techniques. One may then note that the variation in the type of loading that is applied is likely to be as significant as the variation in any one load. It is little use concentrating on the specified load and its variation if, say, torsional oscillation is likely to occur at some operating condition. Whirling may have to be considered. It may well be observed that failures of mechanical parts are often due to loads that were not foreseen, rather than to parts that were incorrectly proportioned to withstand foreseen loads. Statistics is of no help here – as software reliability workers have since discovered when faced with the same problem.

An even more serious objection to statistical method arises from the way in which design is actually carried out. Very little is based on any textbook methods. Probably 90 per cent of actual design is based on the 'what we did last time' technique. The firm or organisation dealing in general mechanical engineering is virtually non-existent nowadays. All specialise to a greater or lesser extent. Consequently, very rarely is a new product designed from first principles. Design is usually the modification of well-established existing designs. The recent success of computer-aided design (CAD), especially when combined with computer-aided manufacture (CADCAM), is that it allows modifications of existing designs to be easily carried out and incorporated into the manufacturing process. Only in isolated cases is CAD an *ab initio* design technique.

The most practised and recognised form of the 'what we did last time' technique is, of course, the use of codes and standards. These sum up the experience of many designers on a national (e.g. BSI) or international (e.g. ISO) scale, rather than relying on the limited experience of a single designer or design team. There may, additionally, be legal implications in the use of codes and standards. However, the seal of approval of such august bodies is only a guarantee that the design conforms to an established practice, which has received widespread approval at the time that the code or standard was set up. I find that all such codes and standards contain very little information on reliability. There are standards dealing with reliability as such, and particularly a comprehensive range dealing with quality control; but I have yet to find, for example, a code or standard that tells me the differences between the design of the product covered by that code or standard for a customer who calls for 99.9 per cent reliability for a given period of use, and that for another customer who calls for, say, 95 per cent reliability in the same product for the same period of use – assuming, that is, that there is a difference in the design, and the customers understand what they are calling for. (If the latter two conditions are not satisfied, I can see no use for quantified reliability!) Nevertheless, codes and standards account for a very large proportion of all design work. It is clear that the choice between statistical and deterministic design is irrelevant to anyone actually using codes and standards for design, i.e. in the vast majority of design situations.

The reader may well think at this stage that I am making a case for conventional design. But that is not so. There can be no doubt that it is the variability in the quantities involved that is responsible for the isolated failure, that unreliability is the occurrence of an unacceptable number of 'isolated' failures, and that variability is best dealt with by well-established statistical techniques: i.e. reliability is a matter of statistics. By contrast, the rationale behind much empirical mechanical design, when it can be identified, simply will not bear scrutiny. So where do we go from here?

In the first place, while the relative merits of a statistical or deterministic approach to design are irrelevant to those using standards or codes, it should most certainly be a major consideration of those responsible for formulating the standards and codes. Secondly, all designs eventually become obsolete. At some time original redesign (i.e. design) is required. What is to be done then? One would suppose that the statistical approach offers the more powerful tool, but design is much more than design for reliability. Whether we reliability workers like it or not, performance must be the prime requirement, to be accompanied most likely by a weight or volume limit and compatibility with other or existing components, with the product to be manufactured using available resources and certainly to a cost limit. Meeting all these requirements must be left to designers with experience of the product concerned, and if they are to use statistical design it should be as compatible as possible with their current methods of thinking, i.e. expressed in terms of the existing design system.

The central objective of this book becomes, then, the provision of a bridge between the conventional and statistical methods of dealing with variability. It is not my intention, or within my competence, to tell experienced designers how to do their job. The practice followed in many textbooks on design, which propound their methods by dealing in turn with the design of various machine elements, such as riveted and other joints, springs, shafts, gears, bearings, cams, flywheels, clutches and so on, will not be adopted in this book. The identification and evaluation of the forces and resulting stress is too well documented to require repetition here. My criticism of many textbooks is that they cover this part of the design process in depth, and often very well, but leave the readers to sort out for themselves what values should be put on the empirical factors. The range of values, when quoted, is often so wide that the design is equally likely to be unreliable on the one hand, or reliable but overweight and hence often inferior in performance and excessively costly on the other. Achieving reliability by over-design is not an economic proposition.

The first step must be to rationalise, as far as possible, the empiricism underlying conventional design, and so in the next chapter we shall examine critically the rationale behind current factors of safety, which lie at the heart of most current methodology. This will form the framework for an appraisal of statistical methods, though we shall first look critically at some of the concepts of conventional statistics.

Before doing this, we need to define the nomenclature to be used. In general, the failure of a mechanical part depends on the 'hostile' nature of the environment that it encounters and its resistive capability. In the majority of cases, the hostile environment is a force giving rise to a stress, defined as force/area, and the resistive capability is one of the usual strength attributes, also measured as a force/area. However, a part subject to any other hostile environment may be said to be 'under stress', where the word 'stress' is now being used more in its everyday meaning rather than in its engineering sense. For example, a high-sided vehicle in a very strong lateral side wind may be said to be under stress. Failure, i.e. overturning, occurs if the restoring moment due to the weight and geometry of the vehicle is less than the disturbing moment (hostile environment) due to the wind. Stress is now being measured as a turning moment. Other examples where the stress is measured in other units spring readily to mind: a corrosive atmosphere is a typical example of some importance. The fit of two parts is another example; in this case load and strength are dimensions. To avoid needless repetition, and yet to maintain complete generality, the word 'stress' will subsequently be used in its widest everyday (or scientific!) sense as a measure of any hostile environment and of the part's corresponding resistive capability. Stress evaluated as force/area is a special case, though the reader with an engineering background may find it convenient to think in terms of this special case most of the time. The hostile environment will be called the 'load' and the resistive capability the 'strength'.

The symbol s will be used to denote stress, and the symbol σ will be reserved for the standard deviation of any quantity, following standard practice. However, the same symbol will be used for the standard deviation of a total population and a sample, relying on the context to make it clear which is intended.

$S(s)$ will be used to denote the strength at any point expressed in terms of the quantity s. Because we shall have to deal with distributed quantities, $S(s)$ is a probability density function (this is discussed in Chapter 3), having a mean value $\bar{S}$ and comprising individual items of strength S_i. If the distribution $S(s)$ has an identifiable minimum it will be denoted by S_{min}. These symbols will be used to denote any strength attribute. For example, $S(s)$ can refer to the ultimate strength, the yield or proof strength, fracture toughness, or any quality necessary to resist a hostile environment. If it is necessary to be more specific in particular cases, special symbols will be defined in the text to which they refer.

Corresponding to $S(s)$, $L(s)$ will be used to denote the probability density function of the stress resulting from applied loads, $\bar{L}$ its mean value, and L_i individual values. If $L(s)$ has an identifiable maximum it will be denoted by L_{max}. A part may be subjected simultaneously to a number of loads, each producing a stress at a given point in the part. These can always be reduced to three principal stresses, and a single equivalent stress derived by one of the maximum principal stress, maximum shear stress (Tresca), or distortion energy (Mises–Hencky) theories. $L(s)$ is this single equivalent value. The value of the stress due to a uniquely defined load in any individual part will, of course, vary throughout that part. Obviously, it is the peak value that promotes failure. It is to be understood therefore that $L(s)$ refers to this peak failure-inducing stress in the following text, and to avoid endless repetition the words 'failure inducing' will not be used, except when it is essential to emphasise the special nature of the stress. Furthermore, it should be understood that the word 'load' is being used in its widest possible context: to paraphrase British Standards somewhat, the load may arise from both externally and internally impressed forces or other impressed phenomena, electrical or mechanical, metallurgical, chemical or biological, temperature or other environmental, dimensional or any other effects, taken either alone or in any combination. They will always be expressed as a stress as described above.

References

Bourne, A.J. (1988) A brief historical review of reliability technology, in *Seminar on Cost Effective Reliability in Mechanical Engineering*, Institution of Mechanical Engineers, London.

Mallagh, C. (1988) The inherent unreliability of reliability data. *Quality and Reliability Engineering International*, **4**, 35–39.

2

Conventional design

It has to be recognised that there is no universally accepted conventional design practice, or a universally accepted nomenclature, when codes and standards are not being used. In this chapter, I shall attempt to describe what seems to me to be the general methodology adopted, and note variations from it as the occasion requires. In the first instance it is advantageous to concentrate exclusively on stress-rupture phenomena. Then design involving wear-out will be examined.

2.1 Stress-rupture failure modes

Stress-rupture is actually a term used mostly in statistical theory, but it is well defined and can apply equally to deterministic work. In a stress-rupture failure the stress gives rise directly and solely to a rupture or fracture or other terminating failure mechanism, and not to some deterioration in the mechanical properties of the part after which it is still possible to operate satisfactorily until the cumulative effects brought on by many applications of load leads to failure. Another way of stating this is that stress-rupture does not ever involve any degradation of the relevant strength attribute (unless, of course, failure occurs). Thus for stress-rupture either the same stress arising in the same individual item will cause failure the first time that it is applied, or else it will never cause failure no matter how many times it is applied. Two premises are made in the conventional treatment of stress-rupture phenomena, though these are nearly always implicit assumptions rather than explicit statements. The first postulates that there is an identifiable maximum load that is imposed on the product. The second postulates that is is possible to identify the strength of the weakest item in the population (to use a statistical term), i.e. among all those items that make up the total manufactured product. It is basic to this approach that the maximum load and the minimum strength are absolute or deterministic. There is no possibility (statistically no probability, i.e. probability = 0) of the maximum load ever being exceeded, or of any item in the product population having a strength less than the minimum identified strength. The deterministic nature of these quantities leads to the description 'deterministic design'. It then follows logically that for stress-rupture failure mechanisms there can be no failures if the minimum strength is greater than the maximum load. Given the premises, this statement is irrefutable. The above criterion is a safeguard against the worst possible case (defined by the premises) that can arise during the operation of the product. It is therefore often referred to as **worst-case design**.

It now remains to identify the maximum load and the minimum strength. Because of the variations in load, the maximum load will in most cases be greater than the nominal load (sometimes a nominal load is taken at what is thought to be the maximum, so the difference could be small or even zero). In any case it is possible to write

$$L_{max} = \phi_L L_{nom} \tag{2.1}$$

where ϕ_L is an empirical factor, L_{max} is the deterministic maximum load, and L_{nom} is the nominal load (also deterministic), which may be defined in the most convenient way for any particular application. In the same way it is possible to write

$$S_{min} = \phi_S S_{nom} \tag{2.2}$$

where ϕ_S is another empirical factor. S_{min} is the minimum strength and S_{nom} is the nominal strength, both being expressed in units relevant to the physical nature of the load. The worst-case design condition can then be written

$$S_{min} > L_{max} \tag{2.3}$$

or substituting from (2.1) and (2.2)

$$\phi_S S_{nom} > \phi_L L_{nom} \tag{2.4}$$

i.e.

$$S_{nom} > \frac{\phi_L}{\phi_S} L_{nom} \tag{2.5}$$

This can be expressed as

$$S_{nom} > \phi L_{nom} \tag{2.6}$$

where $\phi = \phi_L / \phi_S$ is known as the **factor of safety**. Based on the limiting state between survival and failure, the factor of safety is given by

$$\phi = \frac{S_{nom}}{L_{nom}} \tag{2.7}$$

If more than one load is involved, two options are open to the designer. In one option a separate factor can be associated with each load, so that equation (2.4) becomes

$$\phi_S \, S_{nom} > \sum_{j=1}^{j=l} \phi_{Lj} \, L_{nom\,j} \tag{2.8}$$

or

$$S_{nom} > \sum_{j=1}^{j=l} \frac{\phi_{Lj}}{\phi_S} \, L_{nom\,j} \tag{2.9}$$

giving

$$S_{nom} > \sum_{j=1}^{j=l} \phi_j \, L_{nom\,j} \tag{2.10}$$

for l loads. Although written above as a simple summation, the stresses due to the l loads have, of course, to be combined after multiplying by the required factor in accordance with the rules developed in stress analysis disciplines. This option is favoured by, for example, structural engineers. It allows them to use different factors for live and dead loads, to quote one example. The various factors ϕ_j are referred to as **partial factors**. When partial factors are employed, it is not possible to combine them into a single factor of safety, ϕ, as in equations (2.6) and (2.7). The alternative option, used in most mechanical engineering design, combines the stresses due to all the loads before applying any factor. A single factor of safety can then be applied to the combined stress. This enables the form of equation (2.6) to be retained, and involves the use of only a single empirical factor – the factor of safety. It should be noted, however, that only stresses contributing to the same failure mechanism can be added together. Separate failure mechanisms have to be treated separately, each with its own factor of safety.

The single factor is at once the strength and weakness of the technique. On the positive side, only one empirical factor has to be estimated, and this can easily be increased or decreased to maintain correlation as experience is gained. Furthermore, it is very easy to apply during the design process. One has simply to calculate the highest failure-inducing stress due to the nominal load(s) using established stress analysis techniques, and then multiply it by the factor of safety to obtain the worst maximum stress due to those load(s) and from it the required size of the part. On the negative side, it can be seen from equations (2.5) and (2.6) that even in its simplest form the factor of safety is a combination of two other factors, or subfactors, ϕ_L and ϕ_S. It is impossible to assess the contribution of these individual subfactors to the overall factor once the factor itself has been adjusted empirically to correlate design with experience. It is thus impossible to correlate the factor with anything else other than overall experience, and the causes and effects of detailed changes may be incorrectly assessed. More insidiously, the very ease with which the factor can be manipulated leads to what is in effect abuse of its rightful role. For example, an impulsively applied load gives rise to peak stresses that are twice those calculated for a slowly applied load – the usual assumption made in calculating stresses. Rather than apply a factor of 2 as an independent step in the stress calculation, it is often incorporated into the factor of safety. This is particularly attractive if the load is not truly impulsive and the factor is some number less than its theoretical value; indeed, it may vary between individual applications of load and hence, being a distributed quantity, is a good candidate for worst-case treatment. Its inclusion in the overall factor of safety is then an obvious and practical expedient. But the contributions of the different variables to the factor of safety are further obscured as continued empirical adjustments to achieve only overall correlation are made. Yet again, until recently when computers became the universal tool of engineers, stresses were calculated by simplified approximate methods. The factor required to line up the calculated and actual stress was likewise incorporated into the factor of safety. Furthermore, it is well known that the properties of material samples taken from various parts of a finished product can differ as a result of the manufacturing process. This difference may vary from item to item. Rather than use the properties specific to the peak stress location, it is more convenient to factor the nominal strength on a worst-case basis, and then incorporate that into the factor of safety.

And so the list of subfactors grows, until one is not sure what the factor of safety does or does not include – because it is not necessary to declare any subfactor. Any subfactor can be physically ignored but will be automatically included in empirical correlations. This may be thought practically an advantage but is fundamentally, and eventually practically, a source of confusion and ignorance. The factor of safety becomes meaningless, other than as 'the number we used last time'. Finally, there is the practice of increasing the factor a little in difficult cases – 'just to be sure'. This adjustment factor is sometimes called the factor of ignorance, though that would seem to be an admirable description of the factor of safety itself to me! The relationship between the factor of safety and reliability does become somewhat obscure, to say the least, in these circumstances.

The obfuscation referred to in the previous paragraph is compounded by an inability to define the terms in equation (2.4) *et al*. What exactly is nominal? The nominal strength is usually taken as the maker's quoted strength. An approach to makers is met by such vague statements as 'all the material is better than the quoted specification'. Further enquiry leaves one in doubt whether it is the true minimum (this is sometimes claimed, in which case $\phi_s = 1$), or a value that is only very rarely not achieved, how rare never being defined quantitatively. Strength variation will be taken up in more detail in Chapter 3, section 3.2, but, so far as we are concerned, it pales into insignificance compared with nominal load. Shigley (1986) has listed 12 interpretations for nominal load as follows:

1. typical;
2. average;

3. maximum;
4. minimum;
5. expected;
6. rated;
7. usual;
8. they say;
9. limiting;
10. steady;
11. non-steady;
12. estimated.

I think I could add some more, but there is no purpose in doing so; Shigley has made the point quite adequately. Because the definition of 'nominal' is so flexible, and can vary from organisation to organisation involved in the same application, it is clear that universal values for the factor of safety, ϕ, do not exist. When a value of the factor of safety is quoted, one has no idea what it actually means. Shigley seems to accept the current situation as irredeemable, and states that a factor based on experience must be used. He does in fact give some recommended values for ϕ_S, but does not attempt to quote values for ϕ_L. Shigley may well be right in his appraisal of ϕ_L as used in current design methods, but it does leave unresolved the problem of choosing an objective factor of safety, and rules out any comparison of the principles underlying statistical and deterministic design. We cannot therefore let the problem go by default. It can easily be overcome if the nominal values are related to the distribution of values encountered in the field by means of some unequivocal definition. And this is quite possible. Whenever a range of values has to be represented by a single value, the almost universal practice is to quote the mean or average. Sometimes the median or mode is used as a measure of central tendency, but such cases are exceptional. Interpreting the nominal load as the mean value of the load, and the nominal strength as the mean value of the strength, provides a unique definition of the factor of safety as

$$\phi = \frac{\bar{S}}{\bar{L}} \tag{2.11}$$

where $\bar{S}$ is the mean strength and $\bar{L}$ is the mean load. The factor of safety defined in this way is often called the **central factor of safety**, to distinguish it from all others. It may be observed, however, that it is rarely used in design. A more general method of defining the nominal value in terms of statistical parameters will be introduced in Chapter 4, though this is also not used in current practical design.

In the absence of any agreed definition of nominal conditions, it follows that there cannot be any degree of uniformity or standardisation in the basic meaning or values of the factor of safety, though there is no fundamental reason why this should be so. I believe that this lack of precision accounts for much of the mystery that surrounds contemporary design. The stresses in even a complex part can be calculated to an accuracy of a few per cent using modern rigorous techniques, which form part of even a first degree course in engineering (though it requires a computer to do it). Turning to 'real engineering', these accurately calculated stresses are multiplied by factors that can easily differ by 30 per cent, 40 per cent or 50 per cent (or even sometimes more than that) from one practitioner to another, and none can give a rational explanation for the values used. The mind boggles at the inconsistencies and consequent inaccuracies involved. Actual practice makes a mockery of much of engineering science. Nevertheless, I hope the reader will agree that a critically dispassionate appraisal of the design technique reviewed in this chapter would suggest that it is the bookkeeping that is largely responsible for the chaos. The existence of a deterministically identifiable worst case can be challenged, but it can also be defended – a critical appraisal will be made in Chapters 4 and 5 – and if its existence can be substantiated, a rational design procedure can be formulated, which would not be much different from current practice.

Even so, it should be noted that worst-case design can be implemented without resort to factors of safety if the minimum strength and the maximum load could be established by other means. It is also interesting to note that if the difference between either of these quantities and its nominal value is constant – a not unreasonable assumption – then the factor of safety cannot be constant for differing loads: which suggests that current methods could be erroneous in a more fundamental respect. To validate conventional design, it is therefore necessary to validate independently

1. the worst-case concept;
2. the use of factors of safety to calculate the worst-case condition.

The immediate task is therefore to examine thoroughly the worst-case concept. But this can only be done using statistical methods, and so the next chapter is an appreciation of the statistical representation. Meanwhile it is necessary to take up the question of wear-out failure mechanisms, which was deferred at the beginning of this chapter.

2.2 Wear-out failure modes

The role of the factor of safety in design that has to deal with wear processes is much less clear. In a wear process, the load does not cause failure when it is applied, but does cause either immediately or eventually a reduction in some strength attribute of the item to which it is applied: i.e. the load 'damages' the item. Further damage is done each time the load is applied, until failure occurs, either because the part is unable to withstand the same physical load as that which caused the wear damage, e.g. fatigue, or because it is unable to withstand some other load that is totally unrelated: e.g. a part that fails in direct tension after its effective cross-sectional area has been reduced by corrosion. Wear-out differs in two respects from stress-rupture.

First, it is not necessary to eliminate wear as it is to eliminate rupture. It is necessary only to limit the cumulative wear, so that fracture does not occur in the planned or rated life of the product. In fact, it is uneconomic to 'overdesign' a product so that failure is delayed well beyond the rated life.

Second, when a single load is applied, there is no limiting wear condition giving rise to a unique or definable event in the way that fracture is a unique event in the case of stress-rupture. Differing loads just do differing amounts of damage, depending on the damage resistance of the part. This goes on until failure occurs when the cumulative damage exceeds some critical value. It should be noted, however, that in some cases a limiting condition does exist: for example, the well-known fatigue damage curve – the s–N curve – sometimes, but not always, has a minimum stress or limit (the **fatigue limit**). Stresses below this limit do no physical damage. It is thus possible to eliminate all cumulative fatigue damage by treating the fatigue limit as the critical condition in a stress-rupture approach to design. But this is eliminating all wear, and thus ceases to be part of the subject matter of this section. It may also be uneconomic, and this is discussed in Chapter 10. Since, then, there is no quantity corresponding to the minimum strength used in stress-rupture design, it is difficult to see what role the factor ϕ_S has to play in dealing with wear failures.

Likewise, there seems little point in using the factor ϕ_L to estimate a maximum load when we need to know the complete load spectrum in order to assess the cumulative wear damage. This follows immediately from the observation that a load only just above the nominal can do much more damage if it arises frequently than a far higher load that arises only rarely. To overcome this difficulty, some designers use a load histogram, and sum up the damage done by the various loads that make up the histogram using Miner's rule. (Miner's rule is discussed in Chapters 6 and 7.) This is of course primitive statistical methodology. However, if the stresses of a histogram are multiplied by a factor ϕ_L, the histogram is distorted, so that the relative contribution of each load to the cumulative damage is changed. This seems to invalidate, or at least bring into question, the use of ϕ_L in these circumstances.

The difficulty in assessing the role of the factor of safety applied to wear is compounded by the different interpretations placed on the factor by designers. A typical illustration is conventional gear design, which has, at least until quite recently, been based on the standard s–N curve for fatigue fracture and the corresponding curve for pitting. Now any calculation based on the median curve must give the median life; but many good designers have assured me that because a factor of safety has been incorporated into the design method the calculated life will be achieved by about 95 per cent of the gears: i.e. there will be about 5 per cent failures at the specified life, not the 50 per cent that would be derived from the material data on which the design is based. Actually, the figure quoted by various experts varies between 1 per cent and 10 per cent, which is yet another example of the imprecision surrounding contemporary design. In this design methodology a factor on stress is in effect being interpreted as a factor on life. It is difficult to see how this can be justified, and so far as I am aware, no justification has been offered.

Some workers do make a distinction between the roles of the factor of safety in stress-rupture and wear processes, calling the factor necessary to correlate the calculated median life and the minimum field life a **reserve factor**. However, ambiguity still exists, because some designers use this factor in exactly the same way as a factor of safety! Others use the reserve factor on the specified life of the part to obtain a higher life, on which the design is then based. This seems the more rational procedure, so long as the specified life and design life refer to the same cumulative failures.

Yet another method, used frequently for fatigue, involves factoring the stress of the conventional s–N curve to obtain a 'safe' s–N curve on which the design is based, using the known or estimated actual (i.e. not factored) loads. This method will be examined more closely in Chapter 7. It may appear reasonable, but there is no obvious (or even less obvious) fundamental reason for this factoring technique. Some 'safe' s–N curves are derived by special techniques, which have only limited application, and could not be claimed to be general.

The incontrovertible fact is that the amount of wear damage can be estimated only from the actual load (i.e. the stress that it causes) and the actual material properties. The adjustment that a factor of safety is required to make differs so much from one instance to another that it must be regarded as wholly empirical and uniquely specialist. There seems little hope of rationalising such practice. In fairness to those who use such factors, it is worth noting that the forms of wear are so numerous, and nearly every process so imperfectly understood and so markedly non-linear, that empiricism is clearly the first resort of a designer who has many other problems on his or her hands. But we may well ask 'What price reliability?'

The statistical approach to wear will be examined after looking at some of the basic statistics and establishing some understanding of what is involved in connection with its application to stress-rupture processes, but clearly the current design practice with respect to wear cannot be regarded as even qualitatively satisfactory. It is not just the bookkeeping that is in a mess, as may be claimed for stress-rupture; representation of the physical processes is based entirely on empiricism. This is particularly disturbing, because mechanical reliability is dominated by wear lives.

2.3 Achieving reliability by conventional design

How then is reliability achieved in contemporary design? To the extent that the vast majority of design is the copying of previous designs, either indirectly through codes and standards, or directly from existing designs within an organisation, the reliability of the 'new design' can be assumed to be the same as that of the previous designs on which the codes and standards were based, or of the designs that were copied. If that reliability was acceptable, it is assumed that the reliability of the new design will also be acceptable. It cannot be quantified using existing methodology, but a value may exist from field experience of the designs upon which the 'new' one was based, and this is taken to apply to the project in hand. Values are usually contained in generalised or specialised data banks, and may readily be obtained by those contributing to or otherwise having access to such banks. It should be observed that this is not design to a previously specified value – unless that value has been chosen to coincide with field experience (often the case), and contemporary methodology offers no means of departing from these values. Furthermore, not all designs can be of the above nature: otherwise, no progress would ever be made, and we should still be living in the steam age or worse!

The original question should therefore be rephrased so that it becomes: How is reliability achieved in contemporary innovative design? The short answer is, 'It is not'. Designs are carried out using factors of safety based on the designers' experience as already described, with no reference to reliability. Reliability is achieved in a subsequent development process by trial and error – test and fix. In highly developed and sophisticated areas this can be very expensive and time consuming indeed. It is the reason, for example, why few single companies can afford to introduce new aero-engines, and why joint ventures by different consortia or company groupings are the current vogue. Cost of development is also the reason why car manufacturers amalgamated into bigger and bigger firms, so that the high cost of development (and tooling) could be spread over a larger number of their products. This current trend would not be so necessary if most initial innovative designs were more or less acceptable 'off the drawing board', as was often the case in the past when dealing with less sophisticated concepts. It is the heavy dependence on development necessary for much modern machinery that is revealing more and more the inadequacy of contemporary design methodology.

Besides being very expensive and time consuming, development is a blunt-edged tool. It can never cover all eventualities. It also rarely covers more than a fraction of the total life span of a product. For a 'one-off' or limited production it is mostly ineffective. It really is necessary that the initial design be close to the final solution. This ensures a much less costly and time-consuming procurement programme: design is far less expensive, even if it now requires extensive computer facilities in place of the old-style 2B pencil, a rubber and reams of paper. Development is then, as it should only be, fine tuning. Development operates satisfactorily only if problems can be overcome by a series of independent modifications that do not interact. If major interacting modifications are required it is often 'back to square one'. This can be disastrous if it occurs late in a programme. The history of mechanical engineering is littered with examples. It is seen, then, that current design for reliability depends heavily on following past practice, and on test and fix when innovation is involved. The latter technique is no certainty.

It would be unrealistic to expect any design methodology to eliminate the necessity for all development, but it does seem reasonable and desirable that design should require only the backing of confirmatory tests and few costly development modifications. It is by such a criterion that any design methodology must be judged.

References

Shigley, J.E. (1986) *Mechanical Engineering Design*, First Metric Edition, McGraw-Hill International, Singapore.

3
Review of basic statistics

3.1 Statistical distributions

In the statistical approach, the variations in a property due to adventitious causes are described quantitatively and scientifically (so it is claimed) by the **distribution** of that property. In this chapter we shall be making an appraisal of this basic concept of the theory that underlies the statistical approach to design. The mathematical theory is, of course, uncontroversial; but the relationship between the physical world with which we are concerned and the mathematical model needs closer scrutiny.

To take one straightforward example, if we measure the tensile strength of a given material using a number of testpieces made to exactly the same specification, it is well known that we shall not obtain the same result each time. In deterministic work it is assumed that some representative value, such as the mean or average, or perhaps the lowest recorded value, can be used as a measure of the strength of all the testpieces. By contrast, in statistical work the mean or median is used as a measure of central tendency in the customary way, but in addition other parameters, such as the standard deviation, are used to measure the spread or distribution of the values quantitatively. This can be illustrated by means of a typical distribution. Figure 3.1 shows a representation of some distributed data, x. The variate, x, has been used as a universal quantity, which could be any of the physical parameters used in design. Thus x could be the ultimate tensile strength, the hardness, the ductility, the fracture toughness, the fatigue limit, and so on, when representing a material property. It could be a stress, the corrosive nature of an environment, a temperature, a level of vibration, or anything else when representing a load. Specific examples will be discussed later. In Figure 3.1 the number of items (Δq) falling in equal intervals (Δx) of the variate are plotted in standard histogram form against the range of x values to which they apply. A histogram could also be plotted using $\Delta q/\Delta x$ as the ordinate. The quantity $\Delta q/\Delta x$ can be interpreted as a measure of the mean density of the population over the range Δx. Mean population density is here defined as the number of items

21

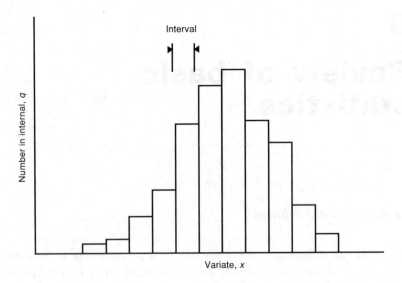

Figure 3.1   A histogram showing variation in some arbitrary quantity.

per unit space (or unit range of the variate) in contrast with, but entirely analogous to, its usual use in connection with humans or animals who inhabit a given space of land. As more and more data become available, the interval Δx can be reduced, so that in the limit $\Delta x \to 0$ and $\Delta q/\Delta x \to dq/dx$. The limiting case of the histogram is a continuous curve, which defines the population density for all values of the variate, x. This is illustrated in Figure 3.2. Taking it a stage further, it is possible to non-dimensionalise this curve by scaling the ordinates so that the area under the curve is equal to unity. The ordinate of such a scaled curve is known as the **probability density**. The curve itself is modelled by the **probability density function** (pdf), denoted by $f(x)$.

The probability that a particular attribute will appear in a population can be evaluated from its probability density function $f(x)$ as follows. The probability of any attribute Z arising in a specified population can be defined as

$$\text{Prob}(Z) = \lim_{Q \to \infty} \left(\frac{q_z}{Q} \right) \tag{3.1}$$

where q_z is the number of items having the attribute Z, and Q is the total number in the population. Taking as a particular attribute those items having values of the variate lying between two specified values, x_1 and x_2, then

$$q_z = \sum_{i=x_1}^{i=x_2} \Delta q_i = \sum_{i=x_1}^{i=x_2} \left(\frac{\Delta q_i}{\Delta x} \right) \Delta x \tag{3.2}$$

where Δq_i is the number of items in the ith interval.

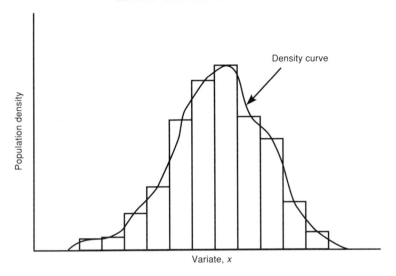

Figure 3.2 Population density histogram and curve.

In the limiting case

$$q_z = \int_{x_1}^{x_2} \left(\frac{dq}{dx}\right) dx \tag{3.3}$$

or when scaled

$$q_z = K \int_{x_1}^{x_2} f(x) \, dx \tag{3.4}$$

where K is the **scaling factor**. Substituting in (3.1):

$$\text{Prob } (x_1 < x < x_2) = \frac{K \int_{x_1}^{x_2} f(x) \, dx}{K \int_{-\infty}^{+\infty} f(x) \, dx} \tag{3.5}$$

Hence

$$\text{Prob } (x_1 < x < x_2) = \int_{x_1}^{x_2} f(x) \, dx \tag{3.6}$$

Equation (3.6) follows from (3.5), because the scale has been chosen so that the integral in the denominator, which is the total area under the probability density function curve, is equal to unity. Equation (3.6) shows that the probability is given by the area under the probability density curve (not by the curve itself).

It follows from equation (3.6) that the probability of observing a value of

x that is less than some value x_0 is

$$\text{Prob } (x < x_0) = \int_{-\infty}^{x_0} f(x) \, dx \tag{3.7}$$

The quantity x_0 can of course be given any value between minus and plus infinity. The right-hand side of equation (3.7) represents the proportion of the total number of items, i.e. a cumulative total, having a value less than x_0. It is thus possible to use x_0 as a variable defining a **cumulative distribution function** (cdf). It is the probability of observing some value of x that is less than x_0.

From equation (3.7),

$$F(x_0) = \int_{-\infty}^{x_0} f(x) \, dx \tag{3.8}$$

where $F(x_0)$ is the cumulative distribution function. It is customary to drop the suffix in writing the cdf, i.e. cdf $= F(x)$. It is then the probability of encountering a value less than x, and has the value of zero when $x = -\infty$ and unity when $x = +\infty$.

As already noted, distributions are usually denoted by two parameters. These are the mean and standard deviation. Sometimes the variance, which is the square of the standard deviation, is used instead of that quantity. The **mean** is a measure of the central tendency, and locates the distribution on the x (variate) axis. For the histogram this is given by

$$\bar{x} = \frac{\sum\limits_{i=1}^{i=I} x_i q_i}{\sum\limits_{i=1}^{i=I} q_i} = \frac{\sum\limits_{i=1}^{i=I} x_i \Delta q_i}{Q} \tag{3.9}$$

where x_i are the individual values of the variate – the mid-interval values of the ith interval – Δq_i is the number of items that appear in the ith interval, and I is the number of intervals. For the continuous (limiting) distribution this becomes

$$\bar{x} = \frac{\int_{-\infty}^{+\infty} x \, f(x) \, dx}{\int_{-\infty}^{+\infty} f(x) \, dx} \tag{3.10}$$

or

$$\bar{x} = \int_{-\infty}^{+\infty} x \, f(x) \, dx \tag{3.11}$$

because the denominator in equation (3.10) equals unity.

The **variance** or **standard deviation** is a measure of the spread of a distribution. The variance is defined as

$$\text{Variance} = \sum_{j=1}^{j=Q} \frac{(x_j - \bar{x})^2}{Q} \tag{3.12}$$

which for the histogram becomes

$$\text{Variance} = \frac{\sum_{i=1}^{i=I} \Delta q_i(x_i - \bar{x})^2}{Q} \tag{3.13}$$

and for the continous distribution

$$\text{Variance} = \int_{-\infty}^{+\infty} (x - \bar{x})^2 f(x) \, dx \tag{3.14}$$

In all cases

$$\text{Standard deviation} = \sqrt{(\text{Variance})} \tag{3.15}$$

It should be noted that if the population is small the denominator in equations (3.9) and (3.13) should be replaced by $Q - 1$. The reasons for this are discussed in textbooks on statistics. If Q is large the difference is insignificant.

As a cruder measure of spread, use can be made of the **range**, which is defined as the difference between the greatest and the smallest observed or perceived values. It has very little significance is mathematical statistics, because it cannot be precisely defined, and nearly always depends on the number of observations. Nevertheless, it is sometimes used by non-statisticians. Because worst-case design is based on the greatest perceived load and the smallest perceived strength, the range is very relevant to that design methodology. It is not possible, as a generality, to convert from range to standard deviation or vice versa, but a kind of correlation is possible, and will be demonstrated later.

Finally, a quantity known as the **coefficient of variation** is often used. It simply expresses the standard deviation as a fraction, or percentage, of the mean value. It is a useful practical way of expressing the standard deviation.

$$\text{Coefficient of variation} = \frac{\sigma}{\bar{x}} \tag{3.16}$$

Some workers quote a signal-to-noise ratio, which is the inverse of the coefficient.

So far, this introduction to statistics has followed a logical sequence common to many textbooks. The significant feature is that no reference has been made, or is required to be made, to any physical property of the variable x. The treatment could be continued by developing the theory in terms of arbitrary distributions. Indeed, some modern statistics does just that. In the classical approach, however, the function $f(x)$ is defined mathematically. This facilitates a mathematical treatment of the subject, which is highly developed. From a practical standpoint, representing test data by a mathematical function can

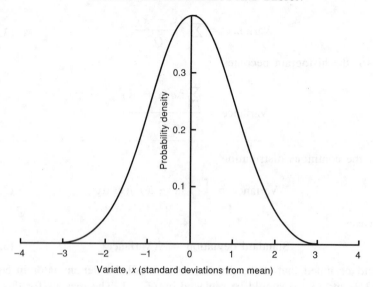

Figure 3.3   Probability density function for Normal distribution (standard deviation = 1).

prove more difficult than it seems. In defining the continuous probability density function it was stated 'as more and more data become available . . .' – a
common enough statement. But what if more and more data do not become
available? It is of course possible to fit arbitrary mathematical functions, such
as a power series or Fourier series, to any data using well-established techniques. But instead of these overt empirical functions, statisticians have preferred other specialised empirical functions. One of the most favoured is the
so-called **Normal distribution**. It would have been far better if, like some
other distributions, it were called after its originator: i.e. called the Gaussian
distribution. Some writers do this, but the description 'Normal' is so widespread in books and tables that it will be retained here. Its probability density
function is given by

$$f(x) = \frac{1}{\sigma\surd(2\pi)} \; e^{-(x-\bar{x})^2/2\sigma^2} \tag{3.17}$$

It fits many cases for which a lot of data are available, and hence is assumed
to be applicable when data are limited. It is necessary only to know the mean
and standard deviation to be able to calculate the probability density for all
values of the variate. Equation (3.17) models the well-known, bell-shaped
distribution shown in Figure 3.3. The cumulative distribution is given by

$$F(x) = \int_{-\infty}^{x} \frac{1}{\sigma\surd(2\pi)} \; e^{-(x-\bar{x})^2/2\sigma^2} \, \mathrm{d}x \tag{3.18}$$

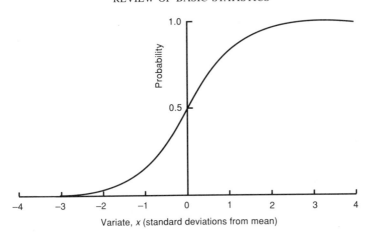

Figure 3.4 Cumulative distribution function for Normal distribution.

Unfortunately, the probability density function cannot be integrated analytically, and values of the cumulative distribution function have to be obtained from tables or computer programs. The cdf for the pdf plotted in Figure 3.3 is given in Figure 3.4. Now the Normal distribution has a number of mathematical properties that make it of special interest to mathematical statisticians, and this no doubt accounts for its pre-eminence in statistical texts. There is no need to doubt these or discuss them here. Additionally, it is often – very often – claimed that many physical measurements are closely approximated by the Normal distribution. The question we must ask is 'How closely?'. Are statisticians and engineers operating under the same assumptions? For example, Moroney (1990), in one of the most popular introductions to statistics, states of the Normal distribution:

> The reader should be on his guard, however, against thinking there is anything abnormal about any observed data which does not follow this law. It is unlikely that any distribution observed in practice follows exactly any of the common distributions used by mathematicians. Careful enough study would in every case bring to light discrepancies.

Quite so. Moroney goes on to say that it is acceptable to use the Normal distribution because 'an accuracy of 1 per cent is ample for most of the real things of life'. Not, I would have thought, if we were dealing with the probability of an aircraft malfunction on landing! It is worth noting here that applied statistics was initially developed (apart from gambling!) largely in connection with biological, including actuarial, purposes, for which an accuracy of 1 per cent or so may be quite adequate. It is totally inadequate for mechanical reliability – particularly if safety is involved – when we may wish to know probabilities of the order of one chance in a million or so. A far cry from 1 per cent. It really is necessary to take stock of the physical situation before accepting established ideas from a totally different field of studies.

To start with, the Normal distribution extends to infinity in either direction, its pdf getting ever closer to the x axis. The complete distribution cannot therefore represent a tensile strength, for example: there is no such thing as a negative strength. (A negative stress is possible, compression being the negative of tension and vice versa, but a negative tensile strength implies an explosion.) Many other properties fall into this category. In order to overcome this difficulty, it is frequently stated that the Normal distribution should be regarded as terminating at some point: three standard deviations ('three sigma') on either side of the mean is often quoted. In fact 99.73 per cent of the area under the pdf falls between these limits. The remainder, 0.27 per cent, lies outside these limits, and is equally divided between the two tails. It follows that 0.135 per cent lies outside a single-sided limit, i.e. roughly one in a thousand lies outside a single-sided three-sigma limit. The assumption that the Normal distribution eventually terminates at three (or any other) sigma from the mean, if valid, does enable us to convert variation expressed as a range limit, into statistical terms. If an experienced person estimates what he or she perceives to be the upper and lower limits, x_u and x_l, of a distributed variate x, together with the average or expected value x_a, then it follows from the three-sigma limit that

$$x_u - x_l = 6\sigma \tag{3.19}$$

or

$$\sigma = \frac{(x_u - x_l)}{6} \tag{3.20}$$

The mean of these estimates is $(x_u + x_l + x_a)/3$. However, the average has already been estimated as x_a. The true mean is therefore taken as the average of these two estimates: that is,

$$\bar{x} = \frac{x_u + 4x_a + x_l}{6} \tag{3.21}$$

It is thus possible to define statistically a distribution that represents the distributed quantity x, initially expressed as a range. Alternative versions of (3.19) and (3.21) can easily be deduced if the limitation is taken at other than the three-sigma points. It is fair to say that the upper and lower limits estimated by reasonably experienced people more often than not do coincide with the three-sigma points in cases for which the true distribution is known by extensive measurement. A little caution is necessary, however.

Using estimates of the upper and lower limits to define the extent of the distribution is putting the statistical method on exactly the same basis as worst-case methods. The upper and lower limits are, by definition, the worst cases. There would seem to be little point in pursuing statistical methods any further. One reason for using statistical techniques is that they offer (or should offer) a method of estimating the exact nature of the complete distribution from the data that are generally available, without having to make empirical estimates. To achieve this, we must be assured that the mathematical distributions that we use accurately represent those parts of the distribution (including the tails) that are relevant to the problem. Glib statements such as 'It is found that the Normal distribution represents many physical phenomena' need to be treated with extreme scepticism. Having been copied from one text to another, they have acquired a totally unjustified aura of omniscience.

The **central limit theorem** is often quoted in justification for the normal distribution. According to this theorem the sum or mean of a number of independent random variables with identical distributions will have an approximate Normal distribution if the number is large enough. Before embracing this argument, one should consider **Pareto's principle**, which is also well supported by experience. This states that in many real-life situations a relatively small number of the independent variables account for the characteristic behaviour of the dependent variable – not the large number required by the central limit theorem. Reasonable doubt must therefore exist concerning the general validity of this argument. On the other hand, if we are lumping together a large number of heterogeneous data sources into one distribution, the argument becomes a very strong one. Thus when we are formulating codes and standards, which must necessarily apply to a large number of independent users, the use of the Normal distribution would most likely be justified.

To cater for the above difficulties there is a whole range of other distributions with which to represent data. Evans *et al.* (1993) list 39 distributions. So there is no shortage of them! These include the log-normal, the beta, the gamma, the Rayleigh, the Weibull, the extreme value and others. But these, too, are nothing more than empirical mathematical equations, having no physical significance, nor any *a priori* justification. It may be apposite here to emphasise that the Weibull distribution is decidedly in this list. Much reliability literature seems to endow it (even the less flexible two-parameter version) with totally unjustified physical properties. The above distributions are sometimes referred to as **parametric distributions** because they are defined by a limited number of parameters: mean and standard deviation for the Normal distribution; locating constant, characteristic life and shaping parameter for the Weibull, and so on. Such distributions are attractive because the parameters defining the distribution can be estimated from limited test data. Extrapolation then provides complete information on the variate. Confidence limits can be placed on the estimated parameters, but only on the assumption that the distribution that best fits the known data also best fits the unknown data. 'Goodness of fit' tests have been devised by statisticians to measure how well a distribution represents the given data, but they are useless for our purpose. They assess the goodness of fit to the known data, not the goodness of fit to the unknown data, which is what we want to know. Choosing one parametric distribution in preference to another is based solely on sophisticated curve-fitting. It would be uncharitable to describe the whole of classical statistics as founded on sophisticated curve-fitting, but it has a grain of truth that can advantageously be kept in mind. The fundamental fact, which this chapter aims to reveal, is that statistical distributions are purely empirical, and have no phenomenological significance whatsoever. The consequences have not been properly addressed in the vast literature or reliability. Hence in the next two sections strength and load distributions will be looked at closely.

3.2 Strength distributions

The question most frequently asked in connection with mechanical reliability studies is 'Where can I get statistical data on materials?' It is a question that cannot be answered. Apart from one or two notable and particular exceptions, generalised statistical data on materials just do not exist, and numerous approaches by reliability workers (including myself) to suppliers of materials encounter a blank wall. No one on the supply side apparently knows or wants to know. One of the most significant facts to me is that of the hundreds of papers I have read on reliability, the many sessions I have attended at reliability conferences and seminars, and the very many books I have read, none attempts a serious appraisal of material variability. Yet this underlies all mechanical reliability. We may also note that there has been no bigger growth area in the university teaching of mechanical engineering than the science of materials, yet reading through the numerous textbooks on the subject there is virtually nothing to be found on the science of material property variation. There are generalised statements of the kind, for example, that electro-slag refining reduces variability. But by how much? As Lord Kelvin once stated, unless you can express a subject quantitatively you do not understand it. There is clearly a problem somewhere. A metallurgist with whom I once discussed the problem suggested that the variability was so great that statistics could not deal with it. I reject that assertion. If statistics is not there to deal with variability, I cannot see any reason for its existence.

Let us look at some of the difficulties. For the distribution of the ultimate tensile strength shown in Figure 3.5 the mean is 594.36 MPa and the standard deviation is 26.76 MPa. The data are reasonably well represented by a Normal distribution having those parameters, as shown in Figure 3.5. The lowest strength given by the three-sigma estimate is 514.08 MPa, which coincides with the lowest value on the histogram. But having determined the parameters for the best fit to the data in Figure 3.5, is it justified to extrapolate? Will a strength less than 514 MPa ever be experienced? It might well be considered that on

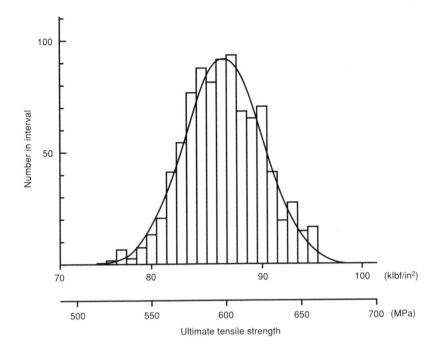

Figure 3.5 Distribution of ultimate tensile strength for SAE 1035 hot-finished carbon steel. Hot-rolled, 1–9 in diameter bars, 913 heats. Data from American Society of Metals (1966).

the basis of this evidence there is some justification for the Normal distribution, and that perhaps some extrapolation is justified. However, Figure 3.6 shows the distribution for the yield strength of the same material, for which the mean is 340.43 MPa and the standard deviation is 25.94 MPa. The lowest three-sigma strength estimate is 262.61 MPa, but the lowest measured yield strength is 275.79 MPa. Are we now prepared to say what is the lowest yield strength to be assumed in design using this material? The difficulty is that in this case the ultimate tensile strength can be represented by a Normal distribution, but the yield strength cannot. The distribution in Figure 3.6 is slightly skewed, and could be better represented by a log-normal or Weibull distribution, but there is no fundamental reason for such distributions. It's just 'the way the cookie crumbled'. Furthermore, it is not possible to say that if testing were continued the yield distribution would become more 'Normal', or indeed whether the distribution of the ultimate strength would become less 'Normal'. Figure 3.7 shows the kind of thing that can happen, with recorded strengths over 5 standard deviations from the mean. It would, of course, be extremely naive to fit parametric distributions to the data in this epiphenomenological manner – though it would be the classical statistical approach –

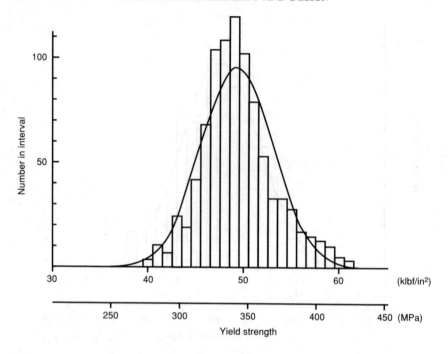

Figure 3.6   Distribution of yield strength for material in Figure 3.5; 899 heats. Data from American Society of Metals (1996).

when it is known that the UTS and yield are causally related through strain-hardening. An essential step in assessing these distributions is therefore to establish that causal relationship.

In his book (1964) Cottrell states (p. 352) that in plastic ruptures

> the external work is done through plastic glide and the thermodynamic condition for this . . . differs from the Griffith condition. Surface energy is usually unimportant, because very much more energy is expended as plastic work along the slip lines. The thermodynamic condition is then simply that the applied stress (allowing for work hardening and plastic constraint effects) should equal the general yield stress.

Initially, yield takes place on a single set of parallel slip planes. Later the deformation becomes random, and 'strong work-hardening sets in as the dislocations moving on intersecting systems mutually entangle one another' (Cottrell, 1964, p. 284). It is reasonable to suppose that the strain-hardening from these random dislocation entanglements is Normally distributed. Let the standard deviation for the whole population of entanglements be σ_g. In any one part there will be a random sample of these dislocation entanglements. Hence the strain-hardening sample means will be distributed with a standard

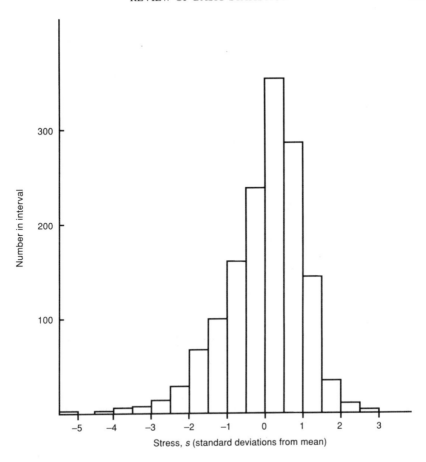

Figure 3.7 Distribution of ultimate tensile strength for a solution-treated and aged titanium alloy, Ti4–A13–Mo10. Mean strength = 200 200 lbf/in² (1380.3 MPa); standard deviation = 7100 lbf/in² (48.95 MPa); 1426 tests. Data from Kecicioglu (1972).

deviation of $\sigma_g/\sqrt{g}$, where g is the number of entanglements in a part. However, g will be very high at the ultimate tensile stress and fracture, i.e. at maximum strain-hardening. Hence the sample means will have near-zero standard deviation: that is, the mean strain-hardenings in each part will tend to the same value. In other words, strain-hardening within a given population is deterministic (to a first approximation). It can thus be concluded from Cottrell's thermodynamic condition that the ultimate tensile stress is uniformly incremented above the yield throughout the whole population. Hence the variation about the mean of the ultimate tensile strength and the yield strength will be substantially the same, and there is a one-on-one ordered relationship between the items in the two distributions. It follows that their

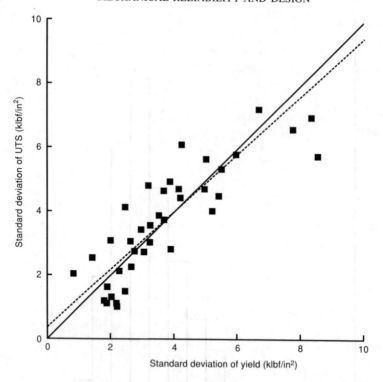

Figure 3.8   The relationship between the standard deviations of the yield and the ultimate tensile strength for steels. ■, American Society of Metals (1966) (test points); —, theoretical relationship; ---, best least-squares straight-line fit.

standard deviations will also be very nearly the same, though of course the means will differ.

There would thus appear to be some conflict between the strength variations deduced theoretically above and that indicated in Figures 3.5 and 3.6. These show different variations, though the standard deviations are substantially the same for yield and UTS. To explore this partial disagreement further, the standard deviations of the yield and UTS for a range of steels have been plotted against each other in Figure 3.8. All the data were obtained from the *American Society of Metals Handbook* (the same source as the data used to produce Figures 3.5 and 3.6), and include every steel for which complete distributions were given. The full line at 45° to the axes defines the theoretical conditions ($\sigma_U = \sigma_Y$). By inspection, the general trend of the test data follows this line. More scientifically, the correlation coefficient is 0.855, which suggests a reasonably strong relationship between the standard deviations of the yield and the ultimate tensile strengths. However, the scatter is considerable, and other factors could be at work. To assess this aspect: suppose the theory were in error. Instead of the strain-hardening being near-deterministic it would then be stochastic. Let it have a standard deviation of σ_{SH}, which is greater than zero. If σ_U is the standard deviation of the ultimate tensile strength, which is obtained by adding the distribution of the strain-hardening to the distribution of the yield strength,

$$\sigma_U = \sqrt{(\sigma_Y^2 + \sigma_{SH}^2)} \qquad (3.23)$$

where σ_Y is the standard deviation of the yield strength.

It follows from this equation that whatever other mechanism of strain hardening may be supposed to operate, the standard deviation of the ultimate tensile strength must always be greater than that of the yield. So it is impossible to have any test points to the right of the full line in Figure 3.8. The only conclusion to be drawn from the fact that actual points do appear in that region is that they do so because of test errors – probably the usual kind of error associated with a tensile test, plus the difficulty in accurately defining the true onset of yield from the standard stress–strain curve (plus possible errors of my own when measuring very small histograms). The most reasonable assumption is that the test error is random, and could be represented by a Normal distribution. With the test points to the right of the full line in Figure 3.8 being used as a measure of test error, it is seen by inspection that the full line is a good estimate of the physical tendency. The statistical best fit (least squares) straight line through the test data is given by

$$\sigma_U = 0.908\sigma_Y + 0.367 \qquad (3.24)$$

It is shown by the dotted line in Figure 3.8. Although not identical with, it is very close to the theoretical line

$$\sigma_U = \sigma_Y \qquad (3.25)$$

over the practical range of standard deviations. Unfortunately, the least-squares straight line supports physically impossible σ_Y, σ_U combinations, and must thus also be considered untrustworthy to some extent. Hence the best relationship, so far as can be ascertained from the standard deviations, would seem to be the theoretical one.

To examine the correlation between actual distributions in closer detail, the ultimate tensile strength distribution and corresponding yield strength distribution for two series of related tests have been produced from the ASM data in Figures 3.9 and 3.10. The former figure relates to varying composition and the latter to pieces taken from varying mass castings (same nominal material). In presenting these figures the ordinates and abscissae have been drawn to the same scale for the two comparison histograms, but the distributions have been translated along the strength axis so that the two means are vertically aligned to facilitate comparison. The UTS is the upper of the two associated histograms. Because of the test errors noted above, a one-to-one correlation between the yield and UTS distributions cannot be expected. On the other hand, the small numbers in some of the samples are no drawback, because

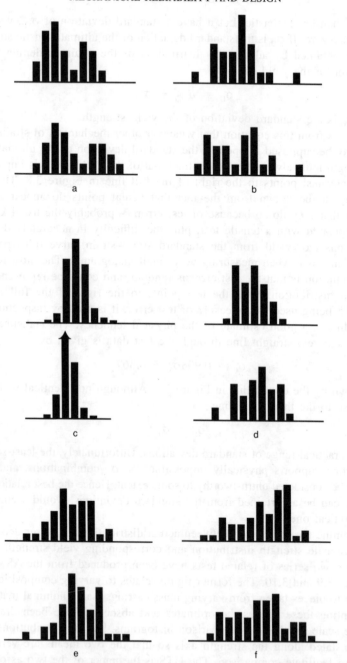

Figure 3.9 Distribution of ultimate tensile strength and yield strength for alloy steels: (a) 25 heats; (b) 25 heats; (c) 200 heats; (d) 40 heats; (e) 50 heats; (f) 50 heats. Mean values of ultimate tensile strength (upper diagram) and yield strength (lower) vertically aligned. Data from American Society of Metals (1966). See reference for further details.

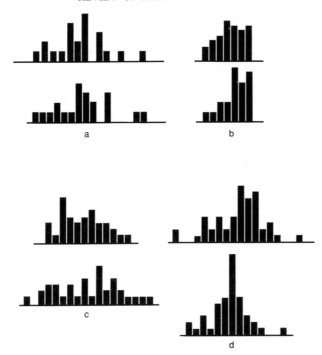

Figure 3.10   Distribution of ultimate tensile strength and yield strength for cast alloy steels from difference mass castings: (a) 22 heats; (b) 50 heats; (c) 37 heats; (d) 44 heats. Mean values of ultimate tensile strength (upper diagram) and yield strength (lower) vertically aligned. Data from American Society of Metals (1966). See reference for further details.

theoretically the UTS is obtained by translating that of the yield to a new mean. (This assumes that the UTS and yield were measured on the same test-pieces.) Within the constraint of the test error it does appear that the general pattern of the yield variation is reflected in that of the UTS. Particular attention is drawn to the observation that when the yield is multi-nodal so too is the UTS; when the tail of the yield is extended or curtailed so too is that of the UTS. It is seen from Figure 3.10 that changes in source mass of the same material bring about corresponding changes in both the yield and UTS distributions. All this is, unfortunately, not quantitative support because of the test errors, but it is strong qualitative evidence that the distributions are behaving in the manner to be expected from theory. The most acceptable assumption is that the theoretical causal variation is the best estimate. It is an important relationship, which will be used extensively in later chapters of this book.

So far as the data in Figure 3.5 and 3.6 are concerned it is not permissible to fit different parametric distributions to both sets of data, and it is not possible to say which of the distributions represents the population behaviour. Clearly a case of 'back to the testing machine'!

It is of course ideally possible by further testing to measure all the data required for design, in which case no reference to parametric distributions need ever be made, except possibly as a mathematical expedient. The question is, how much further testing? Suppose it is desired to record any stress equal to or less than some lowest acceptable stress s_0, where s_0 is defined by its cumulative probability P_0. Tests of varying numbers of pieces can be envisaged. If only one piece were tested then the probability is $(1 - P_0)$ that a stress higher than s_0 would be recorded. If two pieces were tested then the probability that both would have values greater then s_0 is $(1 - P_0)^2$. This is deduced from the product rule on the assumption that the two pieces were truly independent and selected at random from the same population. For θ such pieces the probability that all tests recorded values greater than s_0 would be $(1 - P_0)^\theta$. To be absolutely sure of measuring a stress equal to or less than s_0 an infinite number of tests would be required, i.e. the solution for θ of the equation

$$(1 - P_0)^\theta = 0 \tag{3.26}$$

But suppose a lower confidence (probability) of measuring s_0 were acceptable. Let it equal P_c. Then there is an acceptable probability, $(1 - P_c)$, that all tests record a stress greater than s_0. Equating the probabilities that all the recorded values of s could be greater than s_0 gives

$$(1 - P_0)^\theta = (1 - P_c) \tag{3.27}$$

or, taking logarithms

$$\theta = \frac{\log (1 - P_c)}{\log (1 - P_0)} \tag{3.28}$$

Thus the number of tests required to record a stress equal to or less than that corresponding to a given probability of arising in a population can be calculated at any level of confidence.

As an example, suppose one wishes to be 90 per cent certain ($P_c = 0.9$) of recording a stress that has less than a 0.135 per cent probability of arising in a very large population ($P_0 = 0.001\,35$); then equation (3.26) shows that 1705 tests will be necessary. To be 99 per cent certain would require 3409 tests. A P_0 of 0.001 35 corresponds to the three-sigma point of the Normal distribution, so these figures give some idea of the test effort required. They are directly applicable to Figures 3.5 and 3.6. About twice as many tests as were actually carried out would be required to defined their three-sigma points, even at the lower confidence level. The data of Figure 3.7 are just about acceptable at the 85 per cent confidence level. It is to be noted that this conclusion holds irrespective of the actual distribution, because only the product rule was used in the derivation of equation (3.28). The reader might like to evaluate (3.28) using other practical data to assess the magnitude of the task of defining distributions solely by test.

But perhaps a much more fundamental question should have been asked: What is the population? The population for the topic of this book is the material to be used in the manufacture of the product under design. Furthermore, it is the material in the 'as finished' condition, and this may be very different from the basic material or any testpiece. Even the basic material, although to the same specification, may differ in properties from one supplier to another. In addition to the clear possibility that several suppliers may be involved, there are many other more subtle factors that may be at work. It is not uncommon for manufacturers making several grades of material to transfer a batch that just failed to meet the specification for one grade to a lower grade, or even to supply a higher grade in lieu of a lower one not in stock at the time of the order. It may make financial sense to the supplier, but the stuff received by the customer over a period of time could be a real hotchpotch. The supplier would justify his position by saying that his material met the minimum specification, which is true; but it could make a nonsense of any design methodology. Suppliers often fail to recognise that design for reliability is a continuing trial-and-correct process (test and fix). Spurious variations can corrupt the flow of information both from prototypes and from equipment in the field, invalidating the whole empiricism underlying design for reliability. In statistical terms it violates the IID criterion (independently and indentically distributed). It thus invalidates a lot of statistical analysis of, and deductions from, field failure data.

Variation between suppliers is a well-known and intractable problem. Some quality control authorities, including such well-known ones as Deming in the USA, consider that only one supplier should be used. I regard this as a totally unrealistic proposition, even if it goes a long way to solving the difficulty. It puts the manufacturer completely at the mercy of the sole supplier, both in respect of deliberate financial activities (e.g. those associated with and practised by all monopolies) and with respect to adventitious events (e.g. strikes). Any responsible purchasing department must endeavour to ensure continuity (i.e. reliability) of supply by parallel redundancy, and to avoid common-cause breakdowns will seek as wide a diversity of supply as possible. Wide diversity in supply means wide diversity in strength distribution. Three or four suppliers using discretely different procedures are likely to introduce multinodal distributions rather than the smooth standard ones common in statistical work. It is these that the designer must incorporate into his design methodology. Even so, it may be apposite to note that the Japanese, who were largely motivated by Deming, have a much closer customer–supplier relationship than is customary in the West, and also a high reputation for quality and reliability. The two may not be unconnected.

However, variation in the basic properties of materials is only part of the problem. The manufacturing process itself modifies the properties of the material it is using. It is well known, for example, that testpieces from different parts of a forging or casting can show significantly different properties. This is of course a systematic variation, but the magnitude of the variation from one item to another may still be statistically distributed and thus contribute to the source of finished material variation. Welding, surface finish from machining and other processes, drawing, heat treatment, shot peening, stress releaving, and so on all change the finished material strength from its basic value. None is deterministic. All these are special to any production process, and cannot be included in generalised data. It is not difficult to see why suppliers are reluctant to give precise statistical details of their product when these can, and indeed are most likely to be, distorted during the subsequent manufacturing process.

Theoretically, the effects of various processes on a distribution can be calculated. Thus, if $S_o(s)$ is the original distribution of strength and a process adds $V(s)$ to this, then the final distribution will have a mean given by

$$\bar{S} = \bar{S}_o + \bar{V} \qquad (3.29)$$

and a standard deviation of

$$\sigma = \sqrt{(\sigma_{S_o}^2 + \sigma_V^2)} \qquad (3.30)$$

where S_o and σ_{S_o} are the mean and standard deviation of the original distribution, and $\bar{V}$ and σ_V are the corresponding quantities of the additive process. If the process makes a proportional improvement then

$$\bar{S} = \bar{S}_o \times \bar{V} \qquad (3.31)$$

and

$$\sigma = \sqrt{(\bar{S}_o^2 \, \sigma_V^2 + \bar{V}^2 \sigma_{S_o}^2 + \sigma_{S_o}^2 \, \sigma_V^2)} \qquad (3.32)$$

Substituting representative values shows how the standard deviation of the strength of the final product increases very significantly with the number of processes used in manufacture unless these are very closely controlled.

The practical difficulty is that too many processes, and hence variables, are involved. I do not think my metallurgist friend was correct in saying that the variation in material properties was so great that they could not be represented statistically. What is probably nearer the truth is that the number of independent variables that control the final dependent variation is so great (though only a few may be relevant in each case) that there is no possibility of comprehensive data at any reasonable confidence level being available on all materials of practical significance in the 'as finished' condition. The conclusion is the same, however: no general data on materials are available. This is not to say that no data exist at all. In certain activities, aircraft and nuclear for example, where safety is at a premium, the manufacturing organisations obtain data relating to their particular activities by extensive and continued testing. It is very expensive, and I am not in a position to say how effective it is. Very little is published or generally available as a basis for a disinterested assessment: partly for commercial reasons, and partly because the data apply only to the particular purpose for which they were obtained. Unless such appropriate resources are available to him, the designer can only follow the time-honoured procedure and use his or her own judgement, even for statistical methods aimed at quantified reliability. No more need be said!

3.3 Load distributions

While information on strength distributions may be difficult to come by, that on load distributions, or duty cycles, was virtually non-existent until quite recently. It is amazing to think that designers believed they could produce reliable designs from a knowledge of only the nominal load: particularly when, as we have already seen, no one seems to know what 'nominal' means! But such was the case. It is fair to say, however, that the importance of evaluating the duty cycle before reliable products can be designed, manufactured, and developed has now been fully recognised. Although in the less advanced industries the situation may be no better than ever it was, in reliability-conscious industries more information is being gathered. Much of it is, however, very rudimentary, and there is still a very long way to go. A lot depends on the nature of the activity. In some circumstances the operation of machinery is under statistical control, to a greater or lesser extent. Much of the process industries provide a good example where strict statistical control is often maintained. In these circumstances a reasonable evaluation of the load distribution is possible. However, for general mechanical engineering, as control of the operation becomes more relaxed so it becomes more and more difficult to define the duty cycle.

Determining load distributions to the required accuracy is no easier than defining strength distributions. Equation (3.28) is still valid. If there are a large number of different operators the effort necessary to acquire all the data for all operating circumstances becomes prohibitive. The difficulties associated with three or four different suppliers of material pale into insignificance compared with the difficulties encountered in representing the behaviour of several hundred different operators. The practical expedient is, of course, to concentrate on the most adverse circumstances, but this is a worst-case approach rather than a statistical evaluation of all possible loads.

The problem is made no easier in that collecting data is very expensive indeed. Although modern transducers can be miniaturised, and are fairly robust – which was not the case only a few years ago – they are still difficult to install in most machinery operating 'in the field'. Usually only a very few items of a product, sometimes only one, can be fully instrumented. Data from limited numbers immediately raise questions regarding their representation of actual field practice. This is particularly true if only one item is subject to artificial 'trials'. It is essential that the circumstances to which test data apply should be recorded as fully as possible, and all information conveyed to the designer. Data obtained in this way provide only a rough guide to the kind of distribution that may be expected in actual use and the parameters that control it. They would not provide a detailed distribution upon which quantitative design calculations could be based. In my opinion, superficial data from a large number of items in wide-range actual service are of more immediate design value than data from a comprehensively instrumented 'test vehicle' subjected to artificial trials, though this may be the most convenient first step when no information is available at all.

A further point to which designers must respond is that load distributions can change with time. It is well known that more failures are recorded when a product is first introduced into service than were experienced on prototypes, but that these reduce as the operator learns how to use the equipment – the **learning curve**. This is often absent from later variants. It is not to be confused with 'early life failures'. Less well recognised is that later in life the loading can become more onerous as the operator tries to get more out of available resources. Another age effect is produced by the normal wear and tear of components, which can add to the load on other components in a system. It is the complete distribution of loads over the whole life that must be determined.

For many components and individual parts, the load is determined by the system in which they are installed rather than by the operator. Ideally, these should present less problems and the duty cycle should be well defined, but I find that system engineers are often just as ignorant of the total environment that they are imposing on the subsystems that make up the system, as are operators of the system of the loads that they generate.

To sum up: we must recognise that, as in the case of strength distributions, there are far too many variables for full test coverage, and that statistics can do nothing to define a complete distribution from limited data. We must recognise that load distributions are often far removed from standard distributions: for example, multi-nodal distributions are commonplace for a lot of mechanical equipment. There is often a rational explanation for this, but some mechanical equipment is operated in a most peculiar manner! Finally, and perhaps most significantly, we must recognise that it is far, far too costly and time consuming to measure the distributions comprehensively, except in very special cases. Thus the designers' belief could well be as good an estimate of a load distribution as any other. This is not very reassuring.

There is one other aspect of load distributions that requires attention before proceeding. The histogram form of presentation used earlier would imply that loads are unique and countable. This is not what happens in practice. In reality, a load is often applied and subsequently varied continuously, perhaps over a wide range. Figure 3.11 shows a typical example. What is the load (or loads) in this instance, and how many times has it (or they) been applied?

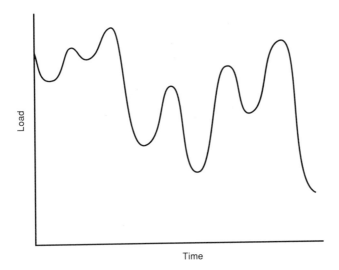

Load

Time

Figure 3.11 Typical unsteady load.

The method usually used to analyse such load records is the so-called **rainflow** technique. The load distribution is plotted vertically with time given a negative value (using standard Cartesian coordinates), and the origin for the abscissa is taken at the mean load. The distribution of Figure 3.11 is plotted in this manner in Figure 3.12. From the appearance of this plot the method is sometimes referred to as the 'pagoda roof' method. The rainflow is imagined to start at the top (origin) of the load history, and is also initiated at the inside of each peak (local load maximum or minimum). The flow is stopped when it reaches flow descending from above, or at a point opposite a peak whose magnitude (unsigned) exceeds the value of the peak from which it started. The point at which the flow is stopped defines the load. If the load is cyclical about the mean, then each flow represents a half cycle, and the other half can be found elsewhere in the record.

The method has been developed almost exclusively by fatigue practitioners. Finney and Denton (1986) point out that the values obtained by the rainflow technique agree closely with those obtained from closed hysteresis loop counts, which has physical significance in the case of fatigue. It can thus be claimed that the method has some physical significance in this instance, but it is not at all obvious that this can be claimed for all failure modes. It is, however, the most convenient and widely used method. It will be assumed that the rainflow technique, or some other if the practitioner prefers, has been applied to the raw data to give a load distribution, of the form described earlier, for use in design.

The distribution used in this book defines the probability that the load will have a given value each time it is applied. Some workers use a distribution

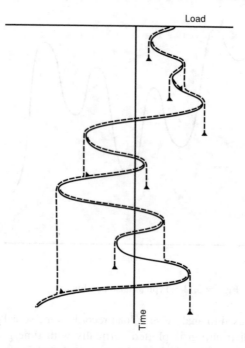

Figure 3.12 Rainflow representation of load in Figure 3.11.

that defines that it will have a given value over a period of time (100 years for some natural phenomena, for example), i.e. over a multiplicity of applications. It may be preferred when only stress-rupture types of failure are involved, but is too restrictive for general mechanical design purposes, which may also have to deal with many wear processes. It will not be used in this book.

3.4 The use of statistical distributions in reliability theory

The mathematics of statistical reliability theory is, of course, unassailable. It is only when specific distributions or numerical values are inserted into deduced formulae that the doubt and the difficulties discussed in this chapter arise. The reader may well consider that the doubts and difficulties discussed in the preceding two sections are so great as to invalidate any design methodology based on statistics (or any other rational procedure for that matter!). However, even if the numerical representation is not accurate, standard mathematical distributions do bear some resemblance to actual distributions obtained in the field in a good many cases. It is therefore possible to use the mathematical distributions in an investigation to identify the major operating characteristics of a design methodology without being committed to quantification. This is an important option, available irrespective of any quantitative correlation or non-correlation between real distributions and their mathematical model. Because it leads to some useful practical conclusions, it is the basic assumption underlying the initial treatment in the subsequent chapters of this book.

3.5 Statistical interpretation of reliability terminology

Before proceeding, it is necessary to specify how reliability and its associated parameters are to be expressed statistically. This is particularly necessary because there is no universally agreed terminology. In this book the British Standards definition of reliability has been adopted. It will be denoted by $R(t)$, and is defined as the ability of a product to perform a required function under stated conditions for a stated period of time. Ability will be measured as the probability of no failure. It will be observed that the BS definition makes no reference to maintenance: i.e. $R(t)$ is the probability of no failure in the absence of maintenance. However, it is obvious that a potential failure can be averted by maintenance, and would not then occur in the field. It is thus necessary to introduce an operational reliability, denoted by $R_{op}(t)$, to take this into account. It is defined as the ability of a product, be it the most simple component or the most complicated plant or equipment, to perform its required function under stated conditions of use and maintenance for a stated period of time. Ability will be measured as the probability of no breakdown. The *Oxford English Dictionary* definition of failure and breakdown will be adopted. 'Fail' is to become exhausted, come to an end, run short, die out. British Standards define 'failure' as the termination of the ability to perform a required function. 'Break down' is to rupture union or continuity, to disrupt, to stop for a time. According to these definitions a failure cannot be repairable; a breakdown is repairable.

It is postulated that parts either fail or do not fail: i.e. there is no intermediate state. If degradation of the part's ability to perform its required functions under the stated conditions of use does take place, then it is assumed that there is some limiting degradation before which the part is fully acceptable, but after which it is unacceptable. If $F(t)$ is the probability of failure at time t

$$R(t) + F(t) = 1 \tag{3.33}$$

or

$$R(t)\% + F(t)\% = 100\% \tag{3.34}$$

In equations (3.33) and (3.34) both $R(t)$ and $F(t)$ are shown explicitly as functions of time, t. Very often this is assumed to be 'understood', so that (3.33) becomes

$$R + F = 1 \qquad (3.35)$$

The time, t, may be measured in normal temporal units (i.e. hours, days, years etc.), or it may be measured in any other convenient units: as a distance, for example, when referring to road vehicle usage. That is, it may be expressed in any units that measure how long the part or machine is under stress. It is merely convenient common practice to use the word 'time' and hence the symbol t to represent them all. In many of the situations to be examined in subsequent chapters, however, reliability will be expressed specifically as a function of the number of load applications, i.e. as $R(n)$. The symbol n is being used to denote time under stress, partly because 'time' is now discontinuous, but more particularly to relate reliability to material properties, which are always expressed in terms of the number of load applications or of load cycles, n.

However, the above definition of 'time' is inadequate by itself. A very simple example will make this clear. I have a common type of wet-shave razor. The blades are moulded into a plastic head and cannot be resharpened. When blades fail (i.e. are no longer sharp enough to give a satisfactory shave) the whole head is discarded. The head is attached to the main body (handle, etc.) by a quick release and fix action mechanism. So when the razor fails to perform its required function, the repair action (replace head) is virtually instantaneous and restores the razor (the machine) to an 'as good as new' condition. The main body itself has effectively an infinite life, becoming outmoded rather than wearing out. This machine thus consists of one maintainable (replaceable) component. Now, starting with a new razor (main body and head) it was found that the first ten head replacements had to be made 13, 26, 38, 49, 64, 80, 93, 110, 124 and 137 days after introduction into service. These data can be plotted as in Figure 3.13, in which the cumulative head failures have been plotted against the age of the main body, i.e. the age of the machine. The points fall more or less on a straight line. The failure rate is given by the slope of this line, and is constant (in this case but not always) at a rate of one head failure per 13.7 days: i.e. the user will require a new head about once every fortnight.

However, the data could equally logically have been presented as in Figure 3.14. The failures have here been plotted against the ordered age of the head, i.e. the age of the repairable (replaceable) component at replacement. That age is, of course, given by the differences in adjacent failure times in the above sequence. It is the age that must be considered most relevant to the design of the head, i.e. the life on which the design of the head will be based. The data plotted in Figure 3.14 use exactly the same scales of Figure 3.13, so that a direct comparison is possible between the two methods of plotting. The shape of the curve in Figure 3.14 is very reminiscent of that in Figure 3.4, i.e. a random variation about a mean life; the reader may well like to plot the data on a much larger timescale to bring out this feature, though it will no longer be comparable with Figure 3.13. The failure rate is

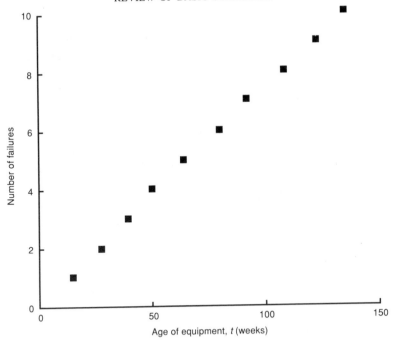

Figure 3.13 Failures versus age of machine.

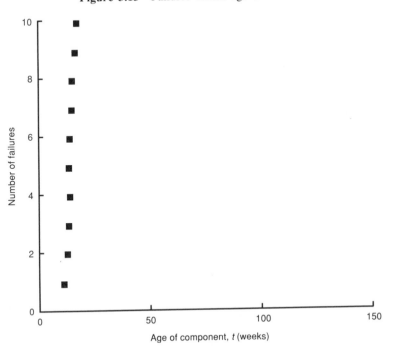

Figure 3.14 Failures versus age of component.

again the slope of the curve, but it is no longer constant – it is indeed zero, or close to it, over most of the component's lifetime, and then increases rapidly.

Both failure rate and failure pattern are totally different in Figures 3.13 and 3.14. This difference stems from the different origins from which time is measured in the two cases: in the first, times to failure are measured consecutively, and in the second concurrently. Logically and mathematically the slopes of the curves in Figures 3.13 and 3.14 can both be called failure rates, because the countable event is identical (a head failure) in both cases, and the incremental time is exactly the same: but to use the phrase 'failure rate' for both is obviously a recipe for confusion and disaster. Yet this is what is done! The fact is that the words 'failure rate' are the easiest to use verbally, so each practitioner uses them for the phenomenon in which he or she is most interested (and claims that the other person is wrong!). Such practices are so ingrained that it is now impossible to eradicate them. Of more concern, some practitioners do not seem to recognise that a difference exists at all. The confusion must be recognised. It appears that no organisation is strong enough to decide on a usage for the words 'failure rate' and impose its definition on the reliability world.

To avoid any confusion in this book, the words 'failure rate' will not be used again. The behaviour of repairable and non-repairable products is examined separately in the next two subsections.

3.5.1 Non-repairable parts and components

It may be thought from Figure 3.14 that as the survivors of a non-repairable product reach a higher age the number of failures actually decreases. But this is only so because there are fewer and fewer survivors to fail. To eliminate this feature it is convenient to express the rate at which failures arise as a fraction of the surviving population, giving a quantity, $h(t)$, which is usually (though not always) called the **hazard**. That terminology will be followed in this book. Because the initial population is statistically taken as unity, the surviving population at time t, will be $R(t)$. Hence

$$\text{Hazard} = h(t) = \frac{1}{R(t)} \frac{dF(t)}{dt} \qquad (3.36)$$

If the hazard is expressed as a function of the number of load applications, then equation (3.36) becomes

$$h(n) = \frac{F(n) - F(n-1)}{R(n)} \qquad n > 0 \qquad (3.37)$$

It is emphasised that t or n measures the age of a non-maintained part or component. From equation (3.33)

$$\frac{dR(t)}{dt} + \frac{dF(t)}{dt} = 0 \qquad (3.38)$$

so that the hazard can be expressed solely in terms of reliability as

$$h(t) = - \frac{1}{R(t)} \frac{dR(t)}{dt} \tag{3.39}$$

Alternatively, equation (3.39) can be written as

$$\frac{dR(t)}{R(t)} = - h(t) \, dt \tag{3.40}$$

Integrating from 0 to t:

$$\log R(t) = - \int_0^t h(t) \, dt \tag{3.41}$$

i.e.

$$R(t) = e^{-\int_0^t h(t) \, dt} \tag{3.42}$$

when the initial reliability at $t = 0$ is 100 per cent or 1.

Some texts use a cumulative hazard function $H(t)$, defined as

$$H(t) = \int_0^t h(t) \, dt \tag{3.43}$$

Equation (3.41) can then be written as

$$R(t) = e^{-H(t)} \tag{3.44}$$

If the hazard, $h(t)$, is constant and equal to h, say, then

$$H(t) = h \times t \tag{3.45}$$

and

$$R(t) = e^{-ht} \tag{3.46}$$

$$= \text{probability of no failure} \tag{3.46a}$$

Although reliability is usually the specified target, the cumulative failures, F, can be used almost synonymously, because they are closely connected by equation (3.35); but very often the hazard (particularly when it is constant) is an even more convenient criterion. Hazard will be used extensively in subsequent chapters. The reliability then varies with time according to equation (3.46). Indeed, reliability is only half the design requirement: it has to be achieved for a specific time. This leads to another measure of reliability. If the hazard is zero, or substantially so, over most of the life, but later increases sharply as wear sets in – a not uncommon situation – the design objective might well be to defer that wear failure. A practical design criterion is then the time to reach a specified percentage of failures, after which the part is considered no longer fit for its purpose. It will be written in the form

T_F or N_F, where F is the percentage failures. Thus T_5 is the life to failure of 5 per cent of the initial population, T_{10} the life of 10 per cent, and so on. This notation will be used extensively later in this book in connection with wear. It is often written as B_F in the literature because it was initially introduced in connection with bearing lives.

3.5.2 Repairable components and systems

Most mechanical machines comprise many elemental components, and when one of these components fails the machine is not discarded but is repaired, the failed component being replaced by a new one. The machine does not fail, but is unable to operate until it is repaired. It is thus said to break down (or to go down, or to be down). The rate at which breakdowns arise will be called the **breakdown rate** in this book, though this is not standard practice. Ascher (1991) has proposed that it be called the **rate of occurrence of failures**, on the grounds that, as already noted, any part or component of the broken-down machine that has to be replaced will have failed. However, this phrase is too long for use in everyday speech, and degenerates back into 'failure rate', so many writers now use its acronym **ROCOF** to describe the rate at which failures arise in a maintained machine. It is, of course, identical to breakdown rate. I prefer the latter as having an etymological meaning self-evident to non-reliability specialists, but the reader can use whichever he or she prefers. It is to be noted that both the breakdown rate and ROCOF include the events connected with all the components of a multi-component machine – the figure of 1/10 000 hr quoted earlier for a gas turbine is an example (constant in that case).

The hazards of mechanical components that make up a repairable assembly are unlikely to be constant, if only because they eventually wear out. This can result in a breakdown rate (ROCOF) that is also not constant, but it can result in a constant breakdown rate (ROCOF) as for the gas turbine above – indeed it must do at infinite time, when all the components will have a random age mix. The razor example used earlier in this chapter shows how it can also occur immediately a machine is brought into use. Most theoretical work concentrates on the limiting condition when the rate is constant.

Suppose the breakdown rate is constant and equal to λ. This means that at time t' there will be $\lambda t'$ breakdowns on average. The symbol t' is here being used for the age of a maintained machine to avoid confusion with t, which refers to a non-maintained component. If it is assumed that the breakdowns occur in a random fashion about the mean and are subject to ideal repair, then the number of breakdowns in time t' must be a Poisson random variable with mean given by

$$\mu = \lambda\, t' \tag{3.47}$$

Ideal repair is one that is instantaneous and restores the machine to the as-new condition. The **homogeneous Poisson process** is given by (see any good textbook on statistics)

$$\text{Prob}(W) = \frac{e^{-\mu} \mu^W}{W!} \quad (W = 0, 1, 2 \ldots; \mu > 0) \tag{3.48}$$

where W is the number of breakdowns.

Now the reliability of a repairable system is the probability of no breakdowns, i.e. the probability that $W = 0$. Hence substituting in (3.48) and using (3.47):

$$R(t) = e^{-\lambda t'} \tag{3.49}$$

$$= \text{probability of no breakdown} \tag{3.49a}$$

Although equations (3.44) and (3.49) are very similar they are fundamentally different, as brought out by equations (3.44a) and (3.49a), and the different t's appearing in equations (3.44) and (3.49). This becomes very apparent when components of non-constant hazards make up a machine that has a constant breakdown rate – see Figures 3.13 and 3.14 for example. The equations refer to different events. They should not be confused. The quantity λ is not a hazard. The hazard that appears in equation (3.44) has no significance for a repairable population.

If the breakdowns are random, but the underlying mean rate is not constant, then the **non-homogeneous Poisson process** (NHPP) has to be used. Even so, it would be as well to recognise that actual repair activities are very far from ideal. In the first place, mechanical repair jobs are rarely instantaneous. Indeed, they can take up a significant time interval, as is only too obvious when machine availability is assessed. Very rarely indeed, even when carried out to perfection, do they restore the machine to the as-new condition. Finally, they are rarely carried out to perfection. The Poisson process could well be a better model of M. Poisson's fertile imagination than of the behaviour of mechanical machines or systems. The fact is that so few data are available, and what exist are inconsistent, that any model will fit some of them. The Poisson model is just the simplest.

So far as the user is concerned, it is often more convenient and relevant to use **availability** rather than breakdown rate or ROCOF. This is defined as

$$\text{Availability} = \frac{\text{Up-time}}{\text{Up-time} + \text{Down-time}} \tag{3.50}$$

It expresses the proportion of the total time that the machine is available to the user. It can also be regarded as the probability that it will be available at any time. Its relevance to the user is obvious, but it is of little use to the designer, except inasmuch as it may be a customer requirement. It is of course absolutely essential that maintenance requirements be met at the design stage, but once this is done, the maintenance contribution to the final reliability must be discounted to obtain the design reliability target of parts.

For that reason, except in the penultimate chapter, this book will be concerned with the design of non-repairable parts that make up the complete product. However, nearly all field data are derived from repairable machines. This is inevitable, because virtually all field data are collected by the user, who is interested only in the behaviour of the complete machine. Ultimately this must be true of the designer, but machine reliability is achieved only via every part. Consequently, many field data are of little direct relevance to the designer, and they must be interpreted in the light of the above observations on repairable and non-repairable products. If the raw repairable data are available, hazards can be deduced by using information only up to first failure in each component of the population in the analysis; otherwise, the designer must proceed with caution. What is called 'censoring' of field data can be particularly troublesome. In this common situation, the failures of one component mask the behaviour of another. It also arises when the field data cease to be available before all components have failed, or some have failed before data were collected. In some cases there can be more censored data than data on failures! The analysis of such information is best left to specialists, and then treated with reserve. One must also be wary of data on products subject to on-condition maintenance. Breakdowns or ROCOFs may include only field failures. For example, the aero-engine industry quotes 'the in-flight shut-down rate'. It is obviously the operational criterion, but many parts may have been replaced 'on condition' before actual failure, so that the in-flight shut-down rate does not equal the replacement rate. In these circumstances, a high operational reliability can still require a large number of component replacements. Most of these difficulties in interpretation will not, in fact, be encountered in this book, but because field data or field experience is the bedrock on which all design must ultimately be based, it is essential that every designer finds out what the data he is going to use really do mean.

References

American Society of Metals (1966) *Metals Handbook*, Vol. 1, *Properties and Selection*, 8th edn.

Ascher, J. (1991) Comments on 'Constant failure rate – a paradigm in transition' by James A. McLinn. *Quality & Reliability Engineering International*, 7 (5), 363–364.

Cottrell, A.H. (1964) *The Mechanical Properties of Matter*, John Wiley & Sons, London.

Evans, M., Hastings N. and Peacock B. (1993) *Statistical Distributions*, 2nd edn, John Wiley & Sons, Chichester.

Finney, J.M. and Denton, A.D. (1986) Cycle counting and reconstitution, with application to the aircraft fatigue data analysis system. *Conference on Fa-*

tigue of Engineering Structures and Materials, Vol. 1, Institution of Mechanical Engineers, London, p. 231.

Kecicioglu, D. (1972) Reliability analysis of mechanical components and systems. *Nuclear Engineering and Design*, **19**, 259–290.

Moroney, M.J. (1990) *Facts from Figures*, Penguin Books, Harmondsworth.

4
Statistical design: stress-rupture modes

4.1 Basic statistical design methodology

In statistical design, the failure-inducing stress arising from the load is not assumed to be unique (deterministic) but is allowed to vary according to some distribution, which must of course be known before the design can proceed. It is the distribution that is used in the calculations. Hence we denote the stress due to the load by a probability density function, $L(s)$. In the same way, the resisting strength is taken to be distributed and represented by the probability density function $S(s)$. It, too, must be known before the design can proceed. It is an essential feature of stress-rupture modes, as contrasted with wear modes, that this function remains invariant with time or number of load applications, except for modifications resulting from failures, which have to be deleted from the distribution as they occur. Naturally, the mean strength will be greater than the mean load, and so the two distributions can be plotted on the same graph in the positions shown in Figure 4.1. It should be recalled that the words 'load' and 'strength' are being used in their widest possible meanings, as discussed earlier.

Failure occurs when the stress due to the load on a particular item of the population exceeds its strength; otherwise, that item does not fail, and remains as good as new. Although the mean strength in Figure 4.1 is considerably greater than the mean load, because both quantities are distributed, it is still possible for a weak item to be subject to a stress that is greater than its strength, and this will cause failure. Failures arise within the overlap or interference range of the load and strength distributions. For this reason this model is often called the **load/strength interference model**. It is a well-established model in the general form of Figure 4.1. An alternative mathematical representation based on the bivariate distribution is sometimes used. It is undoubtedly more versatile, but provides no greater physical insight. The representation in Figure 4.1 is so simple and straightforward as to be a virtually incontrovertible representation of stress-rupture behaviour; it will be used repeatedly.

The first step in assessing this model is to evaluate the reliability of any product that it may represent. To do this we concentrate initially on an arbi-

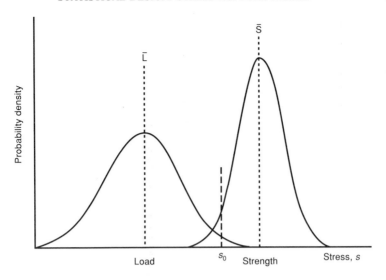

Figure 4.1 The load/strength interference model.

trary strength s_0. The probability of selecting at random from any population an individual item of strength s_0 (strictly lying between s_0 and $s_0 + ds$, as shown by the shaded area in Figure 4.2) is given by

$$\text{Prob } S[s_0 < s < (s_0 + ds)] = S(s_0)ds \qquad (4.1)$$

If the first applied load is greater than s_0, then the item fails and is removed from the population. If the load is less than s_0, the item survives. The probability that the latter condition is met is given from equation (3.6) by

$$\text{Prob } L[s < s_0] = \int_0^{s_0} L(s)ds \qquad (4.2)$$

The probability of selecting an item of strength s_0 and that it survives the application of one load is then the probability of selecting the item from the total population, multiplied by the probability that the load is less than the item's strength. This is the product rule discussed in all textbooks on statistics. Hence

$$\text{Prob of surviving 1 load} = S(s_0)ds \int_0^{s_0} L(s)ds \qquad (4.3)$$

If this item is to survive another application of load, also taken at random from the distribution $L(s)$, then that load too must be less than s_0. Hence again applying the product rule:

$$\text{Prob of surviving 2 loads} = S(s_0)ds \left[\int_0^{s_0} L(s)ds \right]^2 \qquad (4.4)$$

and so on for 3, 4, 5 etc. loads. As a general expression:

$$\text{Prob of surviving } n \text{ loads} = S(s_0)ds \left[\int_0^{s_0} L(s)ds \right]^n \qquad (4.5)$$

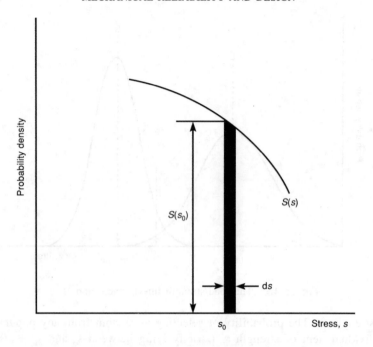

Figure 4.2 Probability of selecting strength s_0.

Now the reliability is measured by the probability of survival of all items of the population, and is obtained by letting s_0 take all values, from 0 to ∞ in the extreme case. Therefore, dropping the suffix o for s in equation (4.5) and adding the probabilities of survival for all possible values of s:

$$\text{Reliability after } n \text{ loads} = \int_0^\infty \left\{ S(s) \left[\int_0^s L(s)\mathrm{d}s \right]^n \right\} \mathrm{d}s \qquad (4.6)$$
$$= R(n)$$

The hazard, $h(n)$, is then given from equations (3.37) and (3.33) by

$$h(n) = \frac{R(n-1) - R(n)}{R(n)} \qquad (4.7)$$

Using equations (4.6) and (4.7) it is possible to calculate the reliability and hence the hazard after any number, n, of load applications. Some typical curves are shown in Figure 4.3.

Following an early life characteristic during which the hazard falls rapidly, the hazard settles down to a constant value, which applies to higher values of n. For design purposes the early life part of the curve can usually be ignored. When it does exist, manufacturers use either burn-in or proof testing to eliminate from the population those items giving rise to these failures before customers receive their supply; or they offer some kind of warranty to cover the

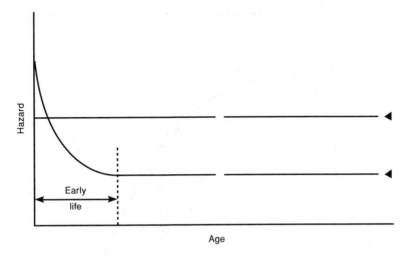

Figure 4.3 Typical hazard–age curves (stress-rupture failure mechanisms). Arrows indicate values plotted in Figure 4.4.

early life period. Burn-in is favoured for electronic equipment. It simply involves the operation of each item for the expected early life period, maybe in a slightly adverse environment. If the item fails during this period it is removed from the population so that the survivors, which the customers receive, can be expected to have the low constant hazard shown in Figure 4.3. Proofing is usually adopted for mechanical products. Completed items off the production line are subject for a shorter period to a somewhat higher load than would be expected in the field. Weak items are again eliminated, and survivors have the constant hazard shown in Figure 4.3. That figure also shows that some products have no early life characteristic, but operate continuously at constant hazard. Proofing or burn-in is ineffective for products following this hazard pattern. For design purposes, then, any early life failures can be ignored and the design based on the constant hazard part of the life history. This is sometimes known as the **random phase**: in practice, it gives way in later life to wear-out – to be discussed in Chapter 6.

The value of the constant hazard depends on the relative shape of the load and strength distributions and their degree of interference. Before discussing numerical values of the hazard, it will be convenient to develop parameters that measure the amount and nature of the interference.

An obvious measure of the extent of the interference between the two distributions is the separation of their means. The amount by which they should be separated depends on the combined spread of the individual distributions. This is measured by a standard deviation equal to the square root of the sum of the squares of the values for the load and strength distributions (see equation 3.30). Hence it is possible to formulate a non-dimensional quantity, known

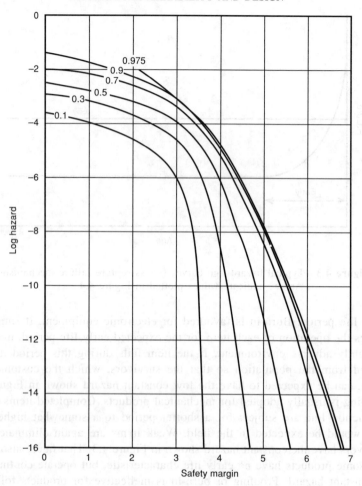

Figure 4.4 Hazard versus safety margin curves for various loading roughnesses
(Weibull-distributed loads and strengths).

as the **safety margin**, with which to measure the separation of the two distributions. It is defined as

$$\text{Safety margin } (SM) = \frac{\overline{S} - \overline{L}}{\sqrt{(\sigma_S^2 + \sigma_L^2)}} \qquad (4.8)$$

where σ_S is the standard deviation of the strength and σ_L is the standard deviation of the load.

A parameter to define the relative shapes of the distributions is also required. In this case the prime quantity we wish to represent is the spread of the load distribution. It can be quantified by its standard deviation, and non-dimensionalised using the standard deviation of the combined load and strength distributions. This parameter is known as the **loading roughness**. It is defined as

$$\text{Loading roughness } (LR) = \frac{\sigma_L}{\sqrt{(\sigma_S^2 + \sigma_L^2)}} \qquad (4.9)$$

We can return now to the design problem. In Figure 4.4 values of the constant hazard for various values of the loading roughness have been plotted against safety margin. Note that a linear–logarithmic form of plot has been used to cover the range of values at adequate resolution. In order to facilitate a numerical solution of equation (4.6) both load and strength were represented by Weibull distributions having a shaping parameter of 3.44. Weibull distributions with that shaping parameter are very similar indeed to, though not identical with, Normal distributions.

Ideally, the curves in Figure 4.4 – or similar ones for other distributions – solve that part of the design problem with which we are immediately concerned. It is necessary only to read off the safety margin corresponding to the hazard derived from the customer's specified reliability at the specified age (using equation 3.46) and the loading roughness determined by the customer's operational requirements. Given the safety margin, the necessary strength of the part, and hence its dimensions, can be calculated. However, practical difficulties do arise, which militate against this simple technique.

The form of each curve in Figure 4.4 can be considered as made up of two major regions, as on the skeleton curve in Figure 4.5. At the low safety margins associated with region 1 the hazard is too high for practical use, and designs falling into this region must be discarded. In these circumstances the main bulk of the distributions are interfering. An early life characteristic is associated with this region. There follows the second region, which includes acceptable hazards. Only the tails of the distributions are interfering in these circumstances, and no early life characteristic is apparent in the fully developed region.

Reference to Figure 4.4 shows that the slope of the curves in region 2 is nearly always greater than about 3. Most often it is very much greater. A value of 3 implies that an error of 0.1 in safety margin gives rise to an error of ratio 2 (i.e. a 100 per cent error) in hazard. This unusual relationship arises from the linear–logarithmic scales. The error obviously increases with slope. The discussion of distributions in Chapter 3 has emphasised that the means of the load and strength distributions, but more particularly their standard deviations, will not be known with great accuracy. Hence we must expect significant differences between the design value of the safety margin and the value subsequently achieved in the field – well in excess of the 0.1 quoted above. Consequent differences between the design and field hazards of one, two, or even three orders of magnitude cannot be discounted. Because the tails of the distributions are interfering, this feature is often known as **tail sensitivity**. In some cases, of course, the field hazard will be less than design

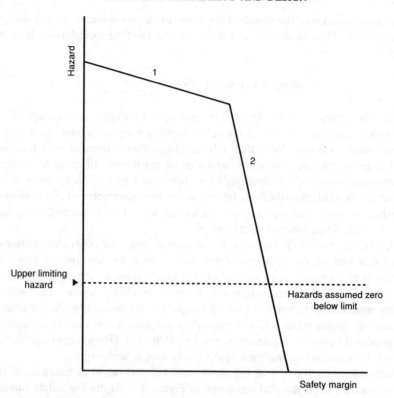

Figure 4.5 Skeleton hazard versus safety margin curve at constant loading roughness.

and the result is still a very satisfied customer. In other cases the hazard will be higher and the result is a dissatisfied (sometimes a very irate) customer. In the second region, then, the hazard may be acceptable but is so sensitive to changes in safety margin and loading roughness that the situation is practically indeterminate. It has been postulated (Carter, 1986) that any design lying in region 2 of Figure 4.5 is totally unacceptable: both because it implies an inability to predict, maybe with disastrous results, the product's behaviour; and also because disparity in operational conditions results in contradictory experience and consequentially poor reputation. Realistically, the bad experience will drive out the good. Design for a specified reliability is not possible.

More recently Taguchi, in his work on experimental design (development), has made the same point in a much more general context. According to his philosophy, designs whose reliabilities (or any other quality criterion) are sensitive to *any* uncontrollable design parameter are unacceptable from a quality point of view. He asserts that designs must be insensitive to all uncontrollable parameters. In these circumstances the designs are said to be **robust**, in the

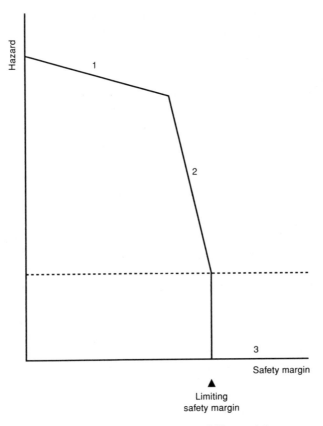

Figure 4.6 A practical version of Figure 4.5.

sense that the uncontrollable parameter can take any random value, within a restricted range, without significantly affecting the resulting quality feature. This is a general design maxim, which is now receiving universal recognition. However, it gives rise to a difficulty in the case under discussion. It is not possible to make use of the insensitive region (i.e. region 1 in Figure 4.5) for design purposes as postulated by Taguchi, because the hazard is unacceptably high in these circumstances.

An alternative approach is necessary. Fortunately, it is possible to turn sensitivity to our advantage; only slight increases in strength attributes, and hence safety margin, are necessary to reduce hazards by many orders of magnitude – in the limit, to hazard values that can be ignored. Designs corresponding to very high sensitivity conditions can also be made robust because all hazards below some negligibly small value (the **limiting hazard**) are effectively zero. The formalised region 2 in Figure 4.5 is in effect terminated at a limiting hazard, so that the whole curve takes the form shown in Figure 4.6, in which safety margins in excess of the limiting value constitute a new region 3. In many instances, the region 2 curve is so steep that it itself forms a termination without any reference to a limiting hazard. In all cases, however, for safety margins in excess of the limiting value the hazards may not be zero, but they are so small that we cannot calculate them with any worthwhile accuracy and, even if we could, the value would have little real significance. For all practical purposes the hazards in region 3 are zero. In legal jargon the failures can be treated as an 'act of God'. It follows that the definition of reliability as a statistical probability has little practical meaning. The design is either unacceptable, if situated in the first two regions; or reliable, in the sense that it is so unlikely to fail that we can assume it will not fail, if situated in region 3. It is interesting to note that such a definition is more in accord with the approach of the common man than with that of the statistician. Nevertheless, it is submitted that there are good sound statistical reasons for taking this position. In order to distinguish this quality from statistically quantifiable reliability, a product that meets these requirements will be described as **intrinsically reliable**. It is postulated that all designs must lie in region 3 of Figure 4.6 and thus be intrinsically reliable.

A straightforward design procedure can be set up to meet this requirement. If the limiting hazard is taken as 10^{-9}, for example, then the safety margin corresponding to this hazard can be read off Figure 4.4 for each value of the loading roughness and plotted as a minimum safety margin–loading roughness curve as in Figure 4.7. Knowing the loading roughness from the customer's requirements, the minimum safety margin can be read from this curve and the part sized as before. There are two points of note in this procedure. First, it allows design only for negligibly small hazards; design for a specified reliability is not possible. Second, there is a range of curves similar to that in Figure 4.7, each corresponding to different load and strength distributions, and possibly to different values of the limiting hazard. There is no 'universal' design curve.

It will be appreciated that intrinsic reliability is obtained in design by adequate separation of the load and strength distributions. 'Adequate' is in fact larger than generally imagined. Figure 4.8 shows the distributed load and strength corresponding to a loading roughness of 0.9 at the minimum separation for intrinsic reliability. This is the configuration that should be kept in mind for practical design – not the formalised version of Figure 4.1, with its high overlap, which is useful only for clarity in the interfering stress range.

4.2 Statistical and worst-case design

Worst-case methodology is based on a minimum strength of a population of products and a maximum load. This implies that the load and strength distributions (for such distributions must exist, even if not recognised explicitly in the methodology) must terminate, or must be considered to terminate, at some

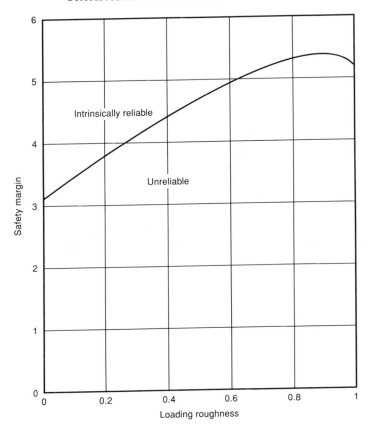

Figure 4.7   Safety margin versus loading roughness to give a hazard of 10^{-9}, a design curve (Weibull-distributed loads and strengths).

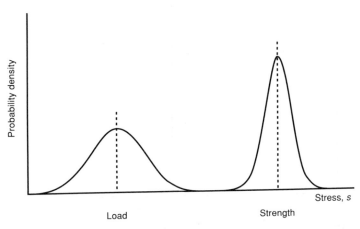

Figure 4.8   The separation of load and strength distributions necessary to achieve intrinsic reliability.

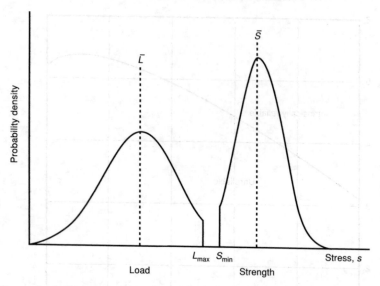

Figure 4.9 The distributions of Figure 4.1 truncated at the worst-case conditions.

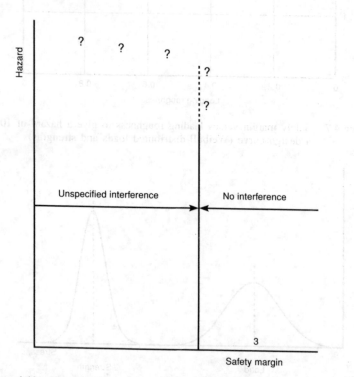

Figure 4.10 Hazard versus safety margin relationship for simple worst-case assumptions; ? denotes unresolved values.

point, as illustrated in Figure 4.9. This should be compared with the conventional representation in Figure 4.1. Worst-case methodology is therefore only a special example of generalised statistical load/strength interference modelling, though this is most often not stated or even recognised – presumably because practical contemporary worst-case methodology does not make use of any statistics. The salient characteristics of worst-case methodology, given its premises, are not difficult to establish.

At very high safety margins the hazard is obviously zero, because there can be no interference between the terminated load and strength distributions. If the safety margin is reduced, the hazard will remain zero so long as the inequality (2.3) is satisfied. When the safety margin is so reduced that the limit state

$$S_{min} = L_{max} \qquad (4.10)$$

is reached, the hazard will jump to a value that is assumed to be unacceptably high. It will of course remain unacceptably high as the safety margin is further reduced. The curve of hazard against safety margin is therefore a vertical line at some critical or limit state point, as shown in Figure 4.10. In the simple worst-case approach the upper value reached by the vertical line is left unresolved. However, a more comprehensive picture may be obtained by using equations (4.6) and (4.7) to evaluate the distributions in Figure 4.9. The resulting curve of hazard against safety margin for a loading roughness of 0.9 is given in Figure 4.11. Curves are given for load termination at four, five and six standard deviations from the mean, the strength being terminated automatically at 3.11 standard deviations in the Weibull distribution. Figure 4.11 shows that apart from a truncation, as there must be when the minimum strength equals the maximum load and the hazard becomes zero, the curve is the same as that in Figure 4.4 over most of the safety margin range: i.e. when significant interference occurs between the distributions, modifying the extreme tail has little effect. When interference ceases, the truncation is abrupt. A formalised linear version of Figure 4.10 would therefore be the same as Figure 4.6. Thus it would appear that there is no qualitative difference between design based on a full statistical analysis (with a hazard limitation) and that based on a statistical interpretation of worst case.

However, it is not necessary to evaluate the complicated statistical equations (4.6) and (4.7) as above in order to estimate the limit state. The limit state condition (4.10) can be written directly even in statistical terms as

$$\bar{S} - k_S \sigma_S = \bar{L} + k_L \sigma_L \qquad (4.11)$$

where k_L and k_S are arbitrary constants that evaluate the difference between the means and the maximum/minimum values in terms of standard deviations. The required strength is then readily estimated as

$$\bar{S} = \bar{L} + k_L \sigma_L + k_S \sigma_S \qquad (4.12)$$

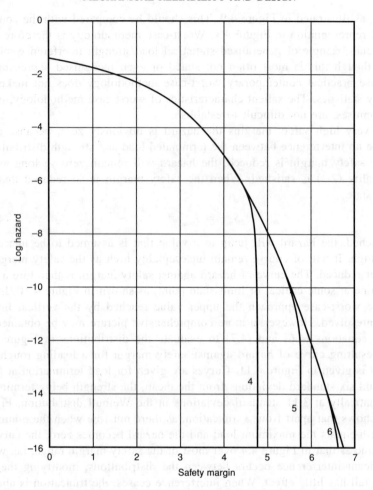

Figure 4.11 Hazard versus safety margin curves for worst-case distributions in Figure 4.9. Numbers on curves = standard deviations of cut-off point from mean.

But how does this estimate compare quantitatively with that obtained from the fully statistical method? A direct comparison is fortunately possible. The loading roughness, LR, is given by

$$LR = \frac{\sigma_L}{\sqrt{(\sigma_S^2 + \sigma_L^2)}} \tag{4.13}$$

so that

$$\sigma_S = \sigma_L \sqrt{\left[\frac{1}{(LR)^2} - 1\right]} \tag{4.14}$$

Hence substituting for σ_S in equation (4.11):

$$\bar{S} - \bar{L} = k_S \, \sigma_L \sqrt{\left[\frac{1}{(LR)^2} - 1 \right]} + k_L \, s_L \tag{4.15}$$

or

$$\frac{\bar{S} - \bar{L}}{\sqrt{(\sigma_S^2 + \sigma_L^2)}} = k_S \frac{\sigma_L}{\sqrt{(\sigma_S^2 + \sigma_L^2)}} \sqrt{\left[\frac{1}{(LR)^2} - 1 \right]} + k_L \frac{\sigma_L}{\sqrt{(\sigma_S^2 + \sigma_L^2)}} \tag{4.16}$$

i.e.

$$SM = k_S \sqrt{[1 - (LR)^2]} + k_L \, LR \tag{4.17}$$

which defines the safety margin at the limit state in terms of the loading roughness and the two constants k_L and k_S, which define the load and strength worst cases. Thus a direct comparison between the limit states defined by negligible hazard and the limit case defined by worst case is readily made, once the limits (i.e. k values) are defined quantitatively.

It is suggested that a negligible hazard can be no greater than 10^{-9}, and this is used in the following comparison. The value is not critical, because the steep slope of the safety margin–hazard curves implies little variation in safety margin at a given loading roughness. The safety margin from Figure 4.7 corresponding to the above value of the hazard is replotted against loading roughness as the full line in Figure 4.12. For worst-case estimation, it was noted in Chapter 3 that the perceived limits of any distribution were given by the 'three sigma' rule. A value of 3 was therefore used for both k_L and k_S. The safety margin calculated by substituting these values in equation (4.17) has been plotted as the lower dotted line in Figure 4.12. Agreement is not good. There is a significant difference between the two curves, which increases as the loading roughness increases: i.e. the statistical conditions for intrinsic reliability do not correspond to the perceived worst-case conditions. The cause is not difficult to find. The three-sigma rule implies that about 1 in 1000 items of a population will lie outside the perceived limits, i.e. violate the assumed worst-case condition. It is suggested that this is not nearly stringent enough a requirement for reliability work. If the worst case is to represent the 'act of God' situation with which intrinsic reliability has been associated, then it should have a chance no greater than about 1 in 1 000 000 of being violated. It must certainly be of this order. A chance of 1 in 1 000 000 corresponds to a k_L of 4.314 and a k_S of 3.04 for the Weibull distribution on which the statistical calculations were based. Using these values in equation (4.17) gives the upper dotted curve in Figure 4.12. It coincides remarkably closely with the statistical curve for intrinsic reliability, except perhaps at unrealistically high loading roughnesses. Bearing in mind tail sensitivity, only small changes in the assumptions would be necessary to achieve even better correlation, but this is considered a pointless line to pursue, for in view of the accuracy

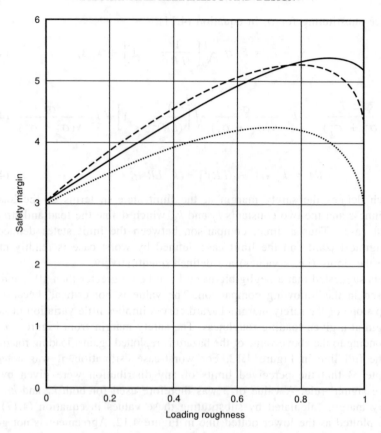

Figure 4.12   Safety margin versus loading roughness design curves for different assumptions: ——, limiting hazard $= 10^{-9}$; ·····, k_s and k_L $= 3$ (perceived values); –––, $k_s = 3.04$ and $k_L = 4.314$ (1 in 1 000 000 condition).

associated with design it would seem that either curve is equally valid. Because neither methodology enables one to design for a prescribed reliability, but only to ensure that failures are so rare that they can be ignored, it may be concluded that simple worst-case design is every bit as good as the full statistical methodology. This should not be surprising. Both methodologies stem from the same basic model, so one should expect to find a great deal of similarity rather than the contrast usually portrayed.

Simple worst-case methodology is far easier to apply, and conforms to contemporary practice. Hence it is reasonable to conclude that it is to be preferred to the full statistical methodology. The latter will not therefore be pursued further, at least so far as stress-rupture failure modes are concerned; but worst-case design will be examined in more detail as a practical methodology. In following this line of investigation there is, however, one important proviso: the choice of k values necessary to achieve correlation shows that the effective worst case to be used in design is *not* the perceived worst case. Other methods must be found to estimate the effective worst case.

Additionally, it must be most strongly emphasised that the agreement between the two methodologies so far demonstrated is nothing more than that. Both may fail to represent the real world that they purport to represent. Immediately, however, it is to be noted that the first objective set out in Chapter 2 has been achieved, and the fundamental validity of worst-case design *vis-à-vis* statistical design has been established. It remains to be seen whether or not worst case can be implemented via factors of safety.

References

Carter, A.D.S. (1986) *Mechanical Reliability*, 2nd edn, Macmillan, Basingstoke.

5

The factor of safety: stress-rupture modes

5.1 Statistical evaluation of the factor of safety

Before evaluating the factor of safety statistically, the reader is reminded of the obscurantism surrounding the factor discussed in Chapter 2. Assumptions will have to be made to remove ill-defined parameters before any quantification can be undertaken. Thus it will be assumed that a single role factor is to be evaluated. In doing so we are not restricting the approach in any way. Secondary functions of the factor can be introduced into the load and strength distributions as described in section 3.2 and treated simultaneously with the primary function, but these secondary functions would have to be defined. Second, it is necessary to define 'nominal'. This is done by identifying any nominal value as lying k_{nom} standard deviations from the mean, where k_{nom} has to be chosen to agree with whatever definition of nominal is being used for the particular application in hand. It is then possible to write

$$L_{nom} = \bar{L} + k_{nomL}\,\sigma_L \tag{5.1}$$

and

$$S_{nom} = \bar{S} - k_{nomS}\,\sigma_S \tag{5.2}$$

from which it follows that

$$\phi_L = \frac{\bar{L} + k_L\,\sigma_L}{\bar{L} + k_{nomL}\,\sigma_L} = \frac{1 + k_L\,\gamma_L}{1 + k_{nomL}\,\gamma_L} \tag{5.3}$$

$$\phi_S = \frac{\bar{S} - k_S\,\sigma_S}{\bar{S} - k_{nomS}\,\sigma_L} = \frac{1 - k_S\,\gamma_S}{1 - k_{nomS}\,\gamma_S} \tag{5.4}$$

$$\phi = \frac{\phi_L}{\phi_S} \tag{5.5}$$

where γ_L and γ_S are the coefficients of variation of the load and strength respectively. It is thus very simple to express the factor of safety in terms of

70

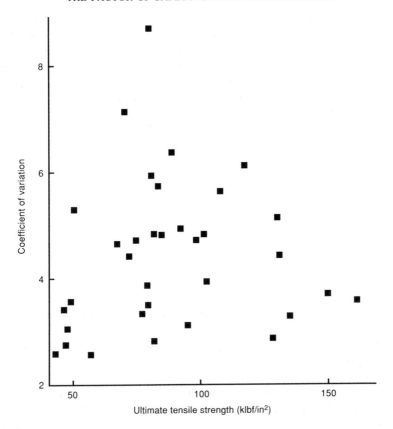

Figure 5.1 Coefficient of variation versus ultimate tensile strength for steels. Data from American Society of Metals (1966).

statistically defined load and strength distributions and use it in a worst-case design, provided of course that the parameters that define the design are completely known.

Now consider a typical design situation in which it is required to redesign an existing satisfactory part using a superior material to reduce size and weight and thus enhance the performance of the machine in which it is fitted. The contemporary method would undoubtedly use the same factor of safety as was used for the original satisfactory part, together with the new material data. How valid is this? Because any reduction in sectional area would apply proportionally to both the standard deviation of the load and its mean, the coefficient of variation of the stress due to the load would remain unchanged. Then equations (5.3), (5.4) and (5.5) show that the coefficient of variation of the material strength must also remain unchanged if the same factor of safety is to apply. But there is no obvious reason why it should. Calling upon test evidence, Figure 5.1 shows the coefficients of variation for the steels used in the analysis leading to Figure 3.7. If the results in Figure 5.1 are anything to judge by, the coefficients of variation are far from constant. As a more exact appraisal, the standard deviations of the ultimate tensile strengths of the steels were assumed to be a linear function of their ultimate tensile strengths. The correlation coefficient is then 0.73, which suggests that there is some correlation between these quantities. The best (least squares) straight-line fit is

$$\sigma_S = 0.0564 \, \bar{S} - 1.044 \tag{5.6}$$

giving

$$\gamma_S = \frac{\sigma_S}{\bar{S}} = 0.0564 - \frac{1.044}{\bar{S}} \tag{5.7a}$$

when $\bar{S}$ is measured in the reported units of klbf/in^2, or

$$\gamma_S = \frac{\sigma_S}{\bar{S}} = 0.564 - \frac{7.198}{\bar{S}} \tag{5.7b}$$

when $\bar{S}$ is measured in MPa.

According to these data the coefficient of variation increases with strength. Intuitively, this is the kind of variation that one would expect, but the scatter is high, suggesting that other factors may be operative. It follows from equation (5.4) that ϕ_S decreases with increased strength, and hence from equation (5.5) that a higher factor of safety is required for the new material. This conflicts with accepted practice, which, as already noted, would retain the same factor. However, equation (5.7) shows that the change in the coefficient of variation for likely changes in the UTS is fairly small, and hence the required change in the factor of safety may well be ignored. For example, if the original design were significantly removed from the limiting condition, retaining the same factor might move the new design closer to that limiting condition without actually transgressing it. The hazard would be negligibly small (effectively zero) for both the original and revised designs. The designer could thus 'get away' with the old factor. If the original design were close to the limiting condition, and the process were repeated, nemesis would of course eventually overtake him!

To obtain a more fundamental understanding of the situation, suppose that the load and strength distributions in Figure 4.1 were moved bodily along the x (stress) axis. Then the ratio between any two arbitrary fixed points, one on the strength distribution and one on the load distribution, would change. If these fixed points are the nominal points, then by equation (2.7) the factor of safety would change. But because the interference between the two distributions would be unchanged, so too would be the reliability. Hence it can be concluded that reliability is not a function of the factor of safety. A similar argument, applied to the truncated distributions of Figure 4.9, indicates that the worst-case condition is also not a function of the factor of safety. Thus the factor of safety is a criterion neither of reliability nor of worst-case conditions, and its use in design is unsound. This may come as a surprise to designers who have used the factor of safety all their lives, but it has been demonstrated in two different ways and must be accepted. It answers the second question raised in Chapter 2 – the factor of safety cannot fundamentally be used to estimate worst-case conditions, though with the caveat that in many circumstances the error may not be of sufficient magnitude to invalidate its use. This is about the most unsatisfactory conclusion that could have been reached!

5.2 Comparison of theory and experience

So far, only different theories have been compared, and it is now really necessary to compare these theories with field experience. It is considered that this is best done by comparing established empirical values of the factor of safety used in conventional design with those calculated from equations (5.3), (5.4) and (5.5). Even so, the obscurantism surrounding empirical factors of safety makes comparing like with like extremely difficult. The author has therefore taken three examples from different fields that he has worked in during his career. The first two are based on his direct knowledge, and the third is one with which he has more of a nodding acquaintance.

5.2.1 Calculation of factors of safety

In each of the examples given below it is considered reasonable to assume, at least *ab initio*, that both the material properties and the applied loads are Normally distributed. Hence k_L and k_S are taken equal to 4.75 in all cases.

Gun tubes (barrels)

The gas maximum chamber pressure in the tube or barrel of a gun must fall within tolerances specified by the Ordnance Board. (Note: maximum is here defined as the highest (average) value obtained when pressure is plotted against shot travel.) Assuming that the OB limits correspond to 3 standard deviations from the mean, the value of $\gamma_L = \sigma_L/\bar{L}$ for the average gun may be taken as 0.013.

The standard deviation of the material strength is about 5 per cent of its mean, so that $\gamma_S = \sigma_S/\bar{S} = 0.05$.

Design is based on the specified maximum gas pressure (i.e. its mean value) and the minimum perceived strength, so that $k_{nomL} = 0$ for the load and $k_{nomS} = 3$ for the strength. Hence using equations (5.3), (5.4) and (5.5):

$$\phi_L = 1 + 4.75 \times 0.013 = 1.062$$

$$\phi_S = \frac{1 - 4.75 \times 0.05}{1 - 3 \times 0.05} = 0.897$$

Calculated value of factor of safety, $\phi = 1.18$.
Value of factor of safety used in design $= 1.2$.

Axial-flow compressor blades

For the type of axial compressor with which the author was concerned, strain gauge measurements showed that the maximum buffeting stress due to wakes and general flow unsteadiness was about equal to the steady gas bending stress. By definition, in this case the maximum recorded stress is the maximum perceived value, i.e. can be taken as 3 standard deviations from the mean, so that $\gamma_L = \sigma_L/\bar{L} = 0.333$.

Buffeting is a cyclic phenomenon, so design must be based on fatigue, but because blade frequencies were so high (measured in kilohertz) the design criterion was that no fatigue damage should be done, i.e. the fatigue limit should not be exceeded. This can therefore be treated as a stress-rupture phenomenon – damage being interpreted as rupture. The standard deviation of the fatigue limit is about 7 per cent of the mean, so $\gamma_S = 0.07$.

Design is based on the steady (i.e. mean) gas bending stress and the minimum perceived strength, so that $k_{nomL} = 0$ for the load and $k_{nomS} = 3$ for the strength. Hence using equations (5.3), (5.4) and (5.5):

$$\phi_L = 1 + 4.75 \times 0.333 = 2.582$$

$$\phi_S = \frac{1 - 4.75 \times 0.07}{1 - 3 \times 0.07} = 0.845$$

Calculated value of factor of safety, $\phi = 3.06$.
Value of factor of safety used in design $= 3$.

Aircraft structures

Variations in load are assumed to have a standard deviation of 25 per cent of the mean design value. Hence $\gamma_L = 0.25$. A representative value of the standard deviation of the strength is 5 per cent of the mean, so $\gamma_S = 0.05$.

Design is based on the maximum perceived load and the minimum perceived strength, so that $k_{nom} = 3$ for both the load and the strength. Hence using equations (5.3), (5.4) and (5.5):

$$\phi_L = \frac{1 + 4.75 \times 0.25}{1 + 3 \times 0.25} = 1.25$$

$$\phi_S = \frac{1 - 4.75 \times 0.05}{1 - 3 \times 0.05} = 0.897$$

Calculated value of factor of safety, $\phi = 1.39$.
Value of factor of safety used in design $= 1.5$.

It should be recorded that this method of design has long been superseded. Nevertheless, it was used for decades and had extensive support. The modern method is based on a fatigue life, and will be discussed fully later under wear.

5.2.2 Appraisal of factors from 5.2.1

The agreement between the calculated and empirical values of the factor of safety is surprisingly good – almost too good to be true, one may think. So let us look at them more closely, starting with the gun. Now the modern gun tube is always autofrettaged. Measurement of the exterior expansion during autofrettage of the unmachined forging is in fact a powerful quality control tool, which could eliminate tubes of inferior material, though the author has been unable to establish whether forgings have been rejected on this score in the past. The numbers are too small for any statistical significance anyway. Nevertheless, such potential quality control could justify a lower value of k_s, which would reduce the calculated value of the factor of safety. Thus if k_s were reduced to 3, which is not an unreasonable assumption in the circumstances envisaged, the calculated value of the factor of safety would be only 1.06. On the other hand, guns are designed using the Lame infinite tube formulae, so the factor should be modified to allow for stress differences arising from constraint at the breach end and pressure discontinuity at the shot-driving face. To this should be added the shot-driving band-engraving pressure, which is also ignored in conventional design. Furthermore the maximum load is highly transient. This has two opposing effects: the fully quasi-static stresses may never be realised during the short time that the maximum pressure is sustained, but the impulsive or shock nature of the loading will increase the stress in the usual way. Presumably all this is incorporated in the empirical factor of safety, because it is not incorporated in the stress calculations. There is an added difficulty: all guns are fired with a proof charge giving 20 per cent increased chamber pressure. At least two old designers have assured the author (though others disagree!) that the factor of safety is used solely to provide for proof firing, i.e. the true factor of safety is unity (cf. 1.06 calculated with autofrettage control). Such claims, of course, beg the question why the proof charge was chosen 20 per cent in excess of the maximum chamber pressure in the first place. This example illustrates how, even in a well-organised design process that is well documented, the exact basis of the empirical factor of safety cannot be fully established. The comparison of calculated and empirical factors has to be taken at its face value.

Turning to compressor blades, the situation is no better. Equating buffeting stress to steady gas bending stress clearly involves strong rounding. It would probably have been more correct to say that buffeting stresses in excess of the gas bending stress had not been recorded. The real value of the buffeting stress was probably somewhat less but largely unknown. The quoted value is just a good safe assumption. Of more significance, the blades were stressed as cantilevers subject to a uniformly distributed (in the engineering not statistical sense) load. While unlaced/unshrouded blades are undoubtedly cantilevers, they are of complicated varying cross-section and twisted, and the real aerodynamic load was certainly not uniformly distibuted even if that was the designer's intention. The relationship between the actual and calculated peak stress was therefore not known, and was incorporated in the factor of safety. The agreement between the calculated and empirical factors of safety could thus be fortuitous, because like is not being compared with like; though the calculated value shows that a much higher factor is necessary for compressor blades than gun tubes, and this does agree with experience.

The aircraft structures provide an interesting example. To start with, the author has been assured that the strengths of many aircraft structures conform more to a Rectangular distribution than the Normal. The only published reference I can find for this interesting feature is in a book for the layman by Gordon (1988) (on pages 327–329). Incidentally, Professor Gordon was involved with early composite structures at the Royal Aircraft Establishment, and hence had every reason to examine closely structural strength variation.

Rectangular distributions make a nonsense of the quoted value for the calculated factor of safety: there can be no justification for taking k_s equal to 4.75. For a true Rectangular distribution the value is $\sqrt{3}$. However, if the Rectangular distribution had a tail it could be very much higher, but no reference is made to tails. Presumably not enough test data were available. No worthwhile comparison of field and calculated factors of safety is possible in such circumstances. It is possible, of course, that if many different structures were considered together a Normal distribution would represent their combined strength distribution (central limit theorem), but the factor of safety is then a global one – not a factor particular to a specific design. Indeed it could be wildly erroneous in specific instances. Nevertheless it may account for the agreement. An unbiased conclusion is that one cannot put too much reliance on the agreement recorded above.

There is a further, more general difficulty in making comparisons. The calculated factors of safety all refer to the limit state. But do the empirical values? Or do they rather satisfy the inequality requirement of (2.3)? It seems highly unlikely that empirical factors of safety could consistently define the limit state. The manner in which they are derived suggests it is more likely that they just meet the inequality requirement. Though inefficient, such designs would be reliable, of course, and this could contribute to the continued use of such empirical factors; but comparisons with calculated values would not then be justified.

In this way doubts can be thrown on all comparisons. The reader must treat the examples given at their face value, and make his or her own comparison. Using examples in which the reader has extensive background knowledge is the only way to put the situation into true perspective. My own view is that calculated values of the factor of safety line up in a general kind of way with the values accepted by custom. In particular, the calculated and empirical values do rank similarly. It is weak, but positive, evidence to justify pursuing the theoretical methodology a little further.

5.3 The factor of safety in perspective

To put the factor of safety into some perspective, it should be applied only to design based on worst-case conditions, i.e. the situation for which it was developed. For convenience in what follows, worst case will be defined as that which is violated on only about 1 in 1 000 000 occasions. If readers disagree with that figure they can insert their own, but it will make little difference to the appraisal unless the value is outrageously different.

There are three possible methods of applying worst case: the worst case can be estimated by empirical factors of safety in the customary manner; it can be deduced from statistical parameters that interpret the design specification; or it can be measured directly. (Intuitive or perceived estimates have already been shown to be invalid.) Of these three possibilities, the last can be ruled out immediately. Even at an absurdly low confidence of 50 per cent, the measurement of the 1 in 1 000 000 chance requires 1 440 000 tests for each distribution (from equation 3.28). To be 95 per cent confident would require 34 000 000! It is all so absurd as to warrant no further consideration. This leaves the first two alternatives as contending methods.

The basic problem underlying all worst-case, or indeed any other, design methodology is the extent to which the distributions have to be separated – see Figure 4.8. It has been shown that the worst-case value should be at least 4.75 standard deviations from the mean for the Normal distribution. For a load whose central distribution is best represented by the Weibull distribution having a shaping parameter of 2 – a common load distribution – the 1 in 1 000 000 point is 6.1 standard deviations from the mean. So, what the designer is trying to do is to estimate the behaviour (variation) of loads and strengths some 4–7 standard deviations from the mean using limited data, usually confined to the central 1 or 2 standard deviations. Scientifically, it is impossible; it has already been made clear in Chapter 3 that statistics is an empirical concept. Extrapolation is invalid. Statistics demands data at the worst-case point in order to proceed, and it is far beyond the resources available to obtain those data. If the statistical methodology for stress-rupture design is to be adopted then k values must be treated as empirical.

Empiricism dominates stress-rupture design, whether using factored nominal values or using statistical defined data. Regrettable and unfortunate though it may be, estimates of the worst case are little more than intelligent guesswork. The usual euphemism is 'guesstimate'! It may not be the kind of thing that one would like associated with one's product, but nothing is to be gained, and much is to be lost, by glossing over this fundamental difficulty. The only point at issue is which method offers the better guess! Looked at from that viewpoint, it has for a long time been my personal belief that a statistically defined worst case has nothing to offer the designer. Conventional design based on conventional factors of safety is every bit as good: better, because it has the backing of so much practical experience. It may be a case of 'better the devil you know than the devil you don't', but that is a very powerful argument when surrounded by a fog of empiricism.

Even so, times and circumstances do change, and some current trends warrant further attention. They fall under four main headings.

5.3.1 Expertise

The expertise that is talked about in textbooks, and to which I myself have customarily turned as a source of quantified factors of safety, is in fact becoming more and more difficult to find on the ground. For reasons associated chiefly with status and salary, few engineers now wish to be permanent members of a design team. 'Management' is the goal. Experience retention in an ever-changing design team is a notorious problem. A factor of safety based on ephemeral experience cannot be as trustworthy as that derived from the more stable experience of the past. The management of any organisation has, of course, the remedy in its own hands if it wishes to retain good designers and a good design team – pay them enough. The cost involved would be more than recouped by the reduced development costs. However, given the lack of technical expertise and the attitude of nearly all contemporary company directors the situation is not likely to improve: indeed it is actually getting worse. In recognition of this, it has been suggested that computer 'expert systems' could replace the human expert. However, expert systems require human experts to provide the input data. Furthermore, computer systems must be logical, and I find it difficult to see how a computer can implement a methodology that is fundamentally flawed. Nevertheless, it would be wise to keep an open mind, and expert systems could yet prove an effective support for conventional design. A more cautious view would be to anticipate a steady decline in the expert knowledge available, and to explore all data retention techniques, including statistical ones.

5.3.2 Statistical distributions

The difficulties involved in ascertaining real statistical distributions have been fully discussed. They are valid, and should in no circumstances be discounted. However, statistical process control is being increasingly applied to manufacture; indeed, some purchasers are already demanding it of their suppliers. This trend will undoubtedly intensify. Statistical process control could go some way to establishing an inherent and sustainable relationship between the main population and the extreme tail. This situation would exist if statistical process control were strictly applied over the whole production process, so that all 'assignable causes' of variation were eliminated. If that were truly achieved, all the remaining sources of variation must be random, and thus there would be good reason to believe that they could be represented by the Normal distribution. It is not easy to achieve – in fact it is very, very, very difficult. The words 'strict' and 'whole' above mean just that. Still, there are grounds for believing that strength distributions may become more tractable in future. The same cannot be said for load distributions. It is true that some operators are adopting statistical process control for their own operations (to reduce scrap and hence costs). Many other operators have little control of the environmental conditions, and many operators are a law unto themselves. It is of course possible to fit governors, load-limiting devices, and so on to control the tail, but opportunities are limited. Rigorous training and the imposition of strict discipline is probably the best form of load quality control, but it can be implemented, and then with some difficulty, only in large organisations. Otherwise, one can see no significant improvement in our knowledge of load distributions, certainly in the near future.

5.3.3 Audit

Design audits are frequently seen as an essential part of any quality assurance programme. But what kind of audit is it that allows values to be multiplied by subjective factors of safety for which there is no, and cannot possibly be any logical, explanation? Factors of safety based on experience can never be measured. The experience on which the factors are based cannot itself be quantified. There is no way of checking beforehand whether the reasoning or intuition leading to the proposed value of the factor has any validity or not: the track record of the expert may generate faith but not knowledge. Current design methodology is opaque. By contrast, in design based on statistically defined inputs all quantities used in calculating the limit condition have a physical significance and cannot be arbitrarily adjusted. If secondary phenomena affect the design, they too have to be specified in the same way as the primary. I believe that this is a valuable salutary discipline. The design methodology is transparent. It is only in such circumstances that an audit is possible.

5.3.4 Stresses

If old methods of manually calculating stresses are replaced by modern computer-based, more exact techniques, the old factors of safety clearly cannot be used – unless the new methods give the same stress as the old, in which case there is little point in using them. New empirical factors are necessary even if the current design methodology is to be retained. It seems doubtful whether wedding exact modern stress analysis technology to an old unsound methodology based on revised factors of safety is the way ahead. The case for starting afresh – as is necessary anyway – with new stress analysis techniques, combined with new statistically defined parameters, is strong.

5.3.5 Conclusions

So where does all this take us? The above four points may be considered to favour the statistical methodology to some extent, but only to a limited extent. In my opinion, the case for worst-case design based on a statistically defined input against worst case based on factors of safety remains 'not proven'. Judgement will consequently be deferred, in this book, until after design for wear has been fully explored: chiefly because, it will be recalled, the use of factors has been called even further into question when applied to the wear processes.

References

American Society of Metals (1960) *Metals Handbook*, Vol. 1, *Properties and Selection*, 8th edn.

Gordon, J.E. (1988) *The New Science of Strong Materials: Or why you don't fall through the floor*, 2nd edn, Penguin Books, Harmondsworth.

6
Wear and wear-out

6.1 Introduction

Wear-out is a much more intractable subject than stress rupture, if only because of the literally hundreds of different wear processes that can lead to failure. In spite of this, a common approach to all wear mechanisms is highly desirable, otherwise design methodology becomes a never-ending topic. A common approach is certainly the target of this book. In principle this is readily achieved, because the methodology of stress-rupture can be carried over to deal with any wear. The only change in the mathematics is a representation of the 'damage' done by each load application; and for those items that survive, the derivation of a new strength with which to assess the reliability of the next load application – and so on.

In purely mathematical terms

$$S(s) = \text{function } [S_o(s), L(s), n] \tag{6.1}$$

where $S_o(s)$ is the initial distribution of strength, $L(s)$ is the load distribution, and n is the number of load applications. The function outlined in equation (6.1) would then be substituted into equation (4.6) in place of the invariant distribution $S(s)$ used for stress-rupture mechanisms. On this basis it is often assumed that at the simplest level equation (6.1) defines a reduction in the mean strength, accompanied perhaps by an increase in its standard deviation.

Successive steps in this process are illustrated at (a), (b), (c) and (d) in Figure 6.1. Intrinsic reliability is achieved by design at (a) and maintained at (b), though with decreasing margin, by virtue of an adequate initial allowance in safety margin. After some time the minimum margin for intrinsic reliability will be reached as at (c). At (d) the degradation has continued further, leading to significant interference between the load and strength distributions and thus a marked increase in the hazard.

The above description of wear would imply that the failures at (d) would be of the same nature as those at (a), were they to occur through an inadequate safety margin. But they are physically different. Failures arising from an initially inadequate safety margin with an invariant strength distribution

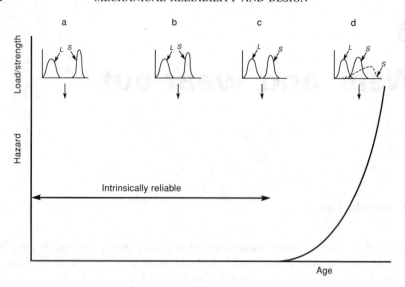

Figure 6.1 A statistical representation of the wear process.

are of an abrupt nature, with no change of state of the item in question prior to failure. Those resulting from strength degradation show physical signs of the degradation, such as erosion, corrosion, creep, elongation, distortion, fraying, leaking, contamination and, in the case of fatigue, cracking prior to actual failure. It is therefore possible to distinguish physically between these types of failure when they occur in the field. In view of this difference one suspects that the above description of the wear process is a little oversimplified. It is only at the simplest level that wear processes modelled by equation (6.1) can justifiably be represented by a steady decrease of strength, as illustrated in Figure 6.1. For example, it is possible that the wear process will change the form of the strength distribution, inasmuch as weak components will wear more rapidly than stronger ones, leading perhaps to the kind of strength distribution shown by the dotted line in Figure 6.1(d).

To obtain a full representation of wear, it is necessary to modify the interference model of Figure 4.1 to include the damage process. The situation at the application of the first load is typically that given in Figure 6.2. As drawn, there is no interference between the load and strength distributions, corresponding to (a) in Figure 6.1. There is, however, interference between the load and damage resistance threshold. The damage threshold is the value of the stress above which damage sets in; below, no damage is done. The damage threshold distribution may represent erosion, corrosion, fatigue or anything else. For the particular case of fatigue the damage threshold is known as the **endurance** or **fatigue limit** (but see Manson (1966) for a possible distinction between the latter two terms, a distinction not observed by most writers).

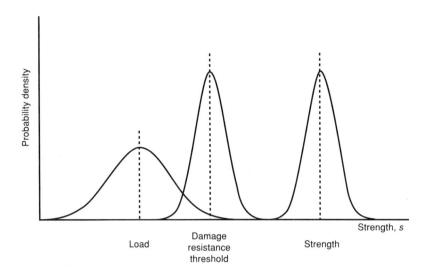

Figure 6.2 Load–damage resistance and load–strength interferences for wear.

It is worth drawing attention even at this stage to the fact that the damage threshold and the strength are not synonymous, as clearly indicated by Figure 6.2. Although there is no interference between the load and strength distributions, there is interference between the load and the damage threshold. The latter may be minimal, of a mild degree as indicated on Figure 6.2, or very much more extensive. The damage will be a function of the difference between any load and the damage threshold, and of the material damage resistance. Damage is cumulative with successive applications of load until a critical value is reached and failure is a consequence. Unfortunately, the damage function is different for every wear process. Furthermore, there are few wear processes for which the function is known, even approximately; complete rooms could be filled with books and papers on fatigue, for example – one of the most common wear processes – and yet the function is not fully resolved. In spite of this, it is highly desirable (even at present essential) that any derivation of the function is based on data that are already available; and the most prolific data on any wear process are those on fatigue. In this case the damage resistance function is represented by the well known s–N curve. It is therefore suggested that this measure of wear damage resistance be adopted as the basis for a statistical approach to all wear processes.

Fatigue specialists may consider that the s–N curve should now be replaced by an evaluation of crack propagation. I have reservations. Although fracture mechanics and crack propagation are now being used to estimate the low-cycle fatigue life of large structures that are initially flawed, the vast majority of mechanical parts are initially flaw free, and designs are still based on the *s–N* curve. In any case, if we wish to start at the more fundamental level, fracture mechanics can always be used to deduce the *s–N* curve. This may not be as easy, however, as is sometimes claimed when a statistical version of the *s–N* curve is required. Hence in the subsequent treatment followed in this book the *s–N* curve will be taken as the starting point.

Another reason for this proposed procedure is that the methodology must also apply to wear processes for which no crack is involved: for it should be noted that an *s–N* curve can be derived (by measurement) for any wear process. Some may not have a 'limit' in the working range that is characteristic of many, but not all, fatigue curves. But one suspects that nearly all wear processes do really have a threshold, i.e. a stress below which no damage is done, though in some cases it may be so small as to have no practical significance.

The damage resistance capability of a material, like any other property, is statistically distributed. Its value at any given value of N, where N is the number of load applications to failure, will be denoted by the probability density function $E_N(s)$. The damage resistance threshold is then $E_\infty(s)$, but because $E_N(s)$ is very often constant over a substantial range of values at high N, i.e. $E_N(s) = E_\infty(s)$ applies to all N values above a critical value, the suffix will be dropped and the **damage resistance threshold** denoted by $E(s)$. This is usually the quoted characteristic property of the material. However, if the threshold $E(s)$ applies only to values of N well outside the working range, it is often convenient to quote the resistance at some other arbitrary chosen value of N (or T if that is the measure used). An arbitrarily defined N could be taken equal to some associated fatigue limit life as a matter of convenience. For example, the fatigue bending and pitting resistance of gears are usually both quoted at 10 000 000 cycles. Above that life the fatigue resistance is constant (the fatigue limit) but the pitting resistance continues to decrease; experts in the subject seem to differ whether a limit is eventually reached or not. Any other reference point is equally valid, of course. Figures 6.3 and 6.4 illustrate the approach.

The *s–N* curve defines the ability of an item to withstand N loads of constant intensity. But real loads are not constant. This leads to the second topic that needs clarification before we can proceed. Virtually all designers make use of Miner's (Miner–Palmgren) law to summate fatigue damage caused by different loads. This law is usually derided (with some justification) by fatigue specialists for being inaccurate or downright misleading at times. However, they have offered no straightforward replacement, and so it continues to be used in design. The theme subsequently developed in this book makes use of Miner's law – also for want of anything better! Furthermore, it will be applied to all other wear processes in addition to fatigue. The law is only a simple linearisation. (Linearisation is a standard technique in scientific work when a more exact treatment becomes intractable, and should thus of itself carry no odium.) Hence if N applications of a constant load cause failure, but

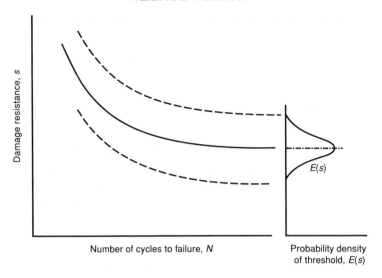

Figure 6.3 *s–N* curves for a distributed damage resistance having a threshold.

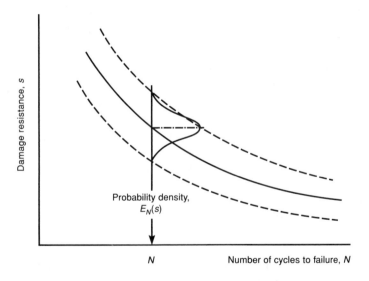

Figure 6.4 *s–N* curves for a distributed damage resistance having no (i.e. zero) threshold.

only n are applied, the damage d done by the n loads is given by

$$d = \frac{\text{damage to failure}}{N} \times n \qquad (6.2)$$

Alternatively

$$d = \frac{n}{N} \qquad (6.3)$$

when d is expressed as a fraction of the damage to cause failure, i.e. $d = 1$ at failure. If i different loads of magnitude L_i are applied n_i times, and N_i is the number to cause failure at load L_i, then the total damage d done by the loads is assumed to be a simple summation of the individual damages, i.e.

$$d = \Sigma\, d_i = \Sigma \frac{n_i}{N_i} \qquad (6.4)$$

and at failure the total damage $d = 1$ so that

$$\Sigma \frac{n_i}{N_i} = 1 \qquad (6.5)$$

This is Miner's law.

6.2 The s–N distribution

The deterministic s–N curve is usually measured by a test at constant stress s until failure occurs after N cycles or applications. Usually two or three tests are carried out at each stress level, though just one is not unusual. Repetitions at various values of s give a number of coordinate (N, s) values for plotting, on either logarithmic or semi-logarithmic axes. A curve is drawn through the middle of the plotted points, usually by eye. Because the distribution of life to failure is rarely Normal, usually having an extended right-hand tail, this construction gives an s–N curve that is a closer approximation to the median than the mean, though it is most often referred to as the mean s–N curve. The full lines in Figures 6.3 and 6.4 are such curves. For fatigue, the value of s below which no testpieces fail after a very large number of load applications is taken as the fatigue limit, which is expressed as a deterministic quantity.

However, it has already been observed that the damage threshold is not deterministic, but is statistically distributed like any other property, as shown on the right-hand side of Fatigue 6.3. It follows that there will be items whose damage threshold is not equal to the 'fatigue limit', i.e. not equal to the median damage threshold. Other such curves of different cumulative failures can therefore

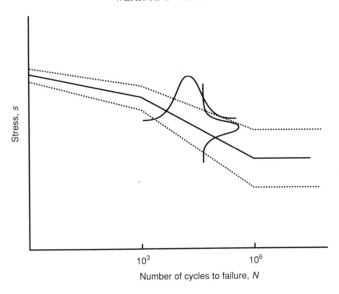

Figure 6.5   A common representation of a distributed s–N relationship found
in the literature, with the distributions of N at constant s and s at
constant N superimposed.

be plotted. The dotted curves in Figures 6.3 and 6.4 are examples, but more
generally one can expect a series of s–N curves following a similar pattern.
For the whole population there will of course be an infinite number of these
curves. Each of the dotted curves is a line of constant cumulative failures. It
will be convenient to refer to these as s–N_F curves, where F is the cumulative
failures expressed as a percentage. In that case the standard s–N curve, which
can be taken to be the median curve and so refers to 50 per cent cumulative
failures, will be designated the s–N_{50} curve. The curve for 5 per cent cumulative failures will be designated the s–N_5 curve, and so on. Each of the curves
will be modelled by a function $\zeta_E(s)$, where E is any value of the damage
threshold taken from the distribution $E(s)$ indicated on Figure 6.3.

The distributed nature of the family of s–N curves can be viewed in two
different ways. The basic variation can be taken either as the distribution of
N at constant s, or with equal validity as the distribution of s at constant N.
Two distributions corresponding to each viewpoint are shown superimposed
on a typical s–N presentation in Figure 6.5. The former is best used for illustrating test data measured at constant stress, but the latter is more relevant to
design for a specified life. The two distributions are of course related. Clearly,
the probability of failure at any (i.e. every) s, N combination must be the
same when evaluated from either of the distributions.

We now run into a difficulty if we try to pursue a rigorous statistical representation of wear. Suppose we have several items, taken from an infinite

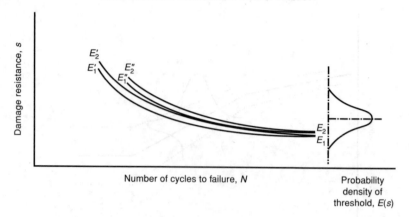

Figure 6.6 Illustrating the crossing of individual s–N curves in a generalised statistical representation.

population, having a damage threshold E_1. They may have the same damage
resistance at high N, but over the full N range each will have its own s–N
curve, which must be taken at random from the whole population of individual s–N curves. In Figure 6.6 two such s–N curves, E_1E_1' and E_1E_1'', have
been plotted for the given E_1. The same would be true for other individual
items whose damage resistance is E_2, for example. These are also plotted in
Figure 6.6. This figure demonstrates that in the most general statistical representation it is possible for individual s–N curves to cross. Physically, this
means that the items that make up the $F\%$ of the s–N_F curve will not be the
same for all values of N. Thus the s–N_F curve does not represent the behaviour of any individual item.

The situation depicted above and illustrated in Figure 6.6 is readily created
if all the individual s–N curves are represented by the ubiquitous power relationship in which the index is statistically distributed. It is the s–N_F curve
that is measured by test. The individual s–N curves cannot be measured in
the absence of a non-destructive fatigue test. However, it is obvious from its
derivation that Miner's rule must apply to the individual (physical) s–N curve
and not to a statistical (abstract) s–N curve. One must ask oneself to what
extent this statistical generalisation represents real material behaviour, and
how Miner is to be applied to a set of s–N_F curves. It is difficult, though not
perhaps impossible, to imagine real s–N curves crossing as in Figure 6.6.
Even if s–N curves are to be treated as purely statistically distributed, one
must still allow for possible covariation between the shape of the curve and
its damage resistance threshold. To account for this it will initially be assumed that the damage resistance distributions at differing values of N can
take any form (to be subsequently determined), but that the ranking of the
individual items within the population is constant and independent of N. Physically, this means that weak items are always weak and stronger items are

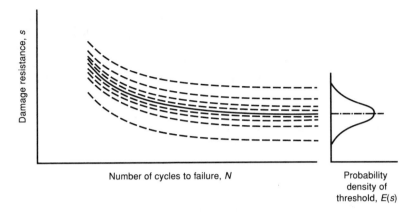

Figure 6.7 Simplified model, in which an individual *s–N* curve is also an *s–N$_F$* curve: ——, median curve (standard *s–N*); – – –, sample of individual curves.

always stronger. Quantitatively, the same items always make up the *F*% of any *s–N$_F$* curve. In that case an *s–N$_F$* curve will always coincide with some individual *s–N* curve in the form of the distribution shown in Figure 6.7, and Miner's rule applies to each *s–N* curve. Later, using modern theory, it will be shown that this condition must be satisfied in a normal fatigue process in high-quality material, though there may be circumstances, which will be identified, when such an assumption is not valid. It must be recognised that there may be some wear process for which it is not valid in any circumstances.

It will be recalled that *s–N* curves can be derived by test for any wear process. However, to obtain more information on the nature of the variation we first turn to the mechanism that has been most studied – fatigue – and to modern theory.

6.2.1 Fatigue

Modern theory gives a good insight into the mechanics of the fatigue process. What follows is not intended to be an introduction to the modern theory. For that purpose the reader is referred, for example, to Miller (1986, 1991) and the many further references quoted therein. What does follow is a recapitulation (based largely on Miller) of those aspects necessary to support the treatment of variation.

Currently, it is believed that the fatal crack responsible for the ultimate fracture is formed at the first application of load. Its source is some surface irregularity or defect. These cracks are known as **microstructurally short cracks**. They grow at a decreasing rate, which can be modelled by the equation

$$\frac{da}{dn} = A \, \Delta\gamma_\rho^\alpha \, (d - a) \tag{6.6}$$

where A and α are material constants, $\Delta\gamma_p$ is the plastic shear strain range, a is the crack length, and d is the distance to the strongest 'barrier'. There may also be weak barriers. The strongest barrier is any feature that stops a microstructurally short crack from propagating further. It may be a grain boundary, an inclusion, or other textural feature. Clearly, from equation (6.6), when the crack length becomes equal to d then da/dn is zero and crack growth will be arrested: at the barrier. There is, however, another form of crack, known as the **physically small crack**, which is not affected by these barriers. It is modelled by the equation

$$\frac{da}{dn} = B \,\Delta\gamma_p^\beta \, a - C \tag{6.7}$$

where B and β are material constants and C is a crack growth threshold. For this type of crack to be initiated and grow, the first term on the right-hand side of equation (6.7) must obviously be greater than C, but if this condition is once met the crack will grow at an ever-increasing rate until fracture occurs. It may be worth recording at this point that the material is behaving plastically over a significant region at the crack tip for both types of crack. Elastic–plastic stress analysis is therefore necessary to evaluate $\Delta\gamma_p$ for any geometric shape and given load.

A physical interpretation of the two crack-growth equations can be represented by the corresponding lines in the plot of da/dn versus a in Figure 6.8. Take first the situation represented by the two upper coarse dotted lines. At the very small imperfections present in a well-finished virgin piece, only a microstructural crack is possible. This grows until its length becomes equal to a_*, the smallest physical crack length at any plastic shear strain, given by

$$a_* = \frac{C}{B \,\Delta\gamma_p^\beta} \tag{6.8}$$

At this stage the crack could be either a microstructurally short crack or a physically small crack, but the crack's behaviour will obviously be determined by the mechanism having the greater growth rate. Hence it continues growing as a microstructurally short crack. However, after the intersection of the two rate of growth lines, the rate of the physically small crack growth becomes greater, and that mechanism takes over. It will thereafter determine the crack growth that is modelled by equation (6.7), though at low stress the crack may become long enough to be modelled by linear elastic fracture mechanics. Failure will occur when the critical stress intensity factor or fracture toughness is reached, or failure may be defined to occur at a specified crack length.

The maximum length of any microstructurally short crack is equal to d: from equation (6.6). In order that a physically small crack can take over it is necessary that

$$d > a_* \tag{6.9}$$

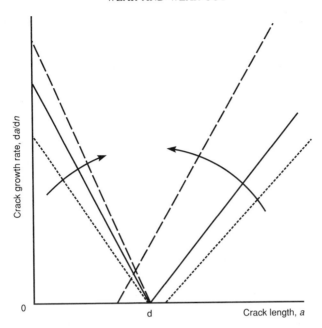

Figure 6.8 Fatigue crack growth: – – –, crack propagation to fracture; ——, 'limit'; ----, crack arrested at d, no fracture; →, increasing stress. From Miller (1991).

If this inequality is not satisfied, as illustrated for example by the lower fine dotted lines in Figure 6.8, it is impossible for any crack to continue growing beyond length d, and the item will consequently have an infinite life. A limiting condition, illustrated by the full lines, which intersect on the abscissa of Figure 6.8, separates the growth to failure and the non-growth behaviours. It corresponds to the **fatigue limit**. According to these modern ideas, the fatigue limit is thus stated to be a limit on the ability of any crack to grow rather than a limit below which cracks cannot form. However, the significance of modern theory for the purpose in hand is that it sees the whole fatigue behaviour as a consequence of crack growth and its modelling by equations (6.6) and (6.7). Although modern theory has not yet addressed variation, that property must also be modelled by those equations if the theory is to stand, and this is the assumption underlying the subsequent treatment in this book.

There is in fact a third equation that applies to the long, low-stress cracks in which the plastic region is confined to a very small area at the crack tip, so that the bulk material behaviour can be modelled by linear elastic fracture mechanics. It is the well-established equation due to Paris, which states

$$\frac{\mathrm{d}a}{\mathrm{d}n} = D \, \Delta K^m \tag{6.10}$$

where ΔK is the stress intensity factor, and D and m are material constants. So far as I am aware, this is not used in initial design, which is still based on empirical factors and codes of practice. It is, however, widely used to estimate the behaviour of existing large structures that are initially flawed, or in any circumstances in which crack lengths can be physically measured. It is thus applied to individual items rather than to a population as a whole, though those items must make up part of some population. The subsequent crack growth can be calculated with confidence by means of equation (6.10), and the time to fracture determined. It is thus at present used more as a condition-monitoring tool than as a design tool, but in this role is an established and trustworthy technique.

As stated earlier, fatigue of initially non-flawed machine parts is modelled by equations (6.6) and (6.7). There are six parameters defining crack growth in these equations, to which must be added the incipient crack length, a_o, at the time that the first load is applied, and the criterion defining failure. If all these are allowed to vary independently and in a random fashion, one is presented with a truly formidable problem. At best, crack behaviour could be handled only by Monte Carlo techniques. Even that is not a contemporary option because the variation, and any covariation, of the parameters is not known. In order to reduce the problem to manageable proportions and derive only a qualitative understanding of the process, some simplification is necessary. Guidance can be obtained from the practices followed in applying the long-established Paris equation (6.10). Although some workers do treat D as a variate (a coefficient of variation of 3 per cent is sometimes quoted, but I have seen no experimental evidence for this), m is always taken as determinate. However, the majority of workers treat both D and m as determinate, so that the variation in life originates solely from the variation in initial crack length. This assumption is essential if the life of an arbitrary item from a population is to be assessed, because otherwise it would be impossible to know what values of D and m applied to the particular item. The results of such estimates are supported by experience. This simple assumption (i.e. that all parameters in the propagation equations can be treated as determinate) will therefore be taken across to the treatment of equations (6.6) and (6.7).

Now deterministic calculations for well-finished pieces in which $a_o \ll d$ show that the incipient crack length has little effect on life. This is due to the very small fraction of the total life spent in the early stages. Figure 6.8 shows that the initial crack growth rate is very high, and hence only a small change in the number of applied load cycles is necessary to counter any variation in the incipient crack length. Thereafter, the growth rate decreases rapidly to near zero, as indicated by Figure 6.8. A very long time (cycles) is spent at crack lengths approaching d, under which condition the growth rate is near zero and large numbers of applied loads are necessary to increment the crack. It is these that very largely account for the total life.

This may be seen more directly from Figure 6.9, in which the crack length

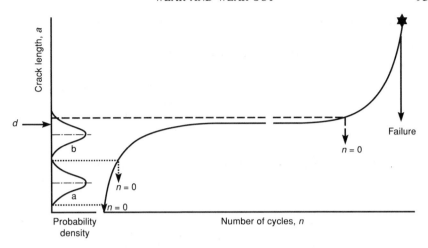

Figure 6.9 Fatigue crack growth at constant plastic shear strain: ·····, extreme values from distribution a; ———, value from distribution b.

has been plotted against the number of cycles from zero crack length to a value at which failure is assumed to occur. Because the parameters in the propagation equations have been assumed invariant this is a unique curve, but different items of the population will enter it at different points, depending on the initial crack length. Thus, referring to the distribution of crack lengths shown at (a) in Figure 6.9, what may be considered the shortest and longest initial cracks enter the regime at the points shown by the dotted lines. On account of the initial very steep slope of the curve, the entry points of the different crack lengths are very close together on the n scale. In fact the difference is negligible in comparison with the number of cycles to failure.

Approaching fracture, the crack is once again growing so fast (see Figures 6.8 and 6.9) that variations in the failure criterion are likewise swamped. Thus if the variation in incipient crack length is not great, so that all incipient cracks are significantly less than the distance to the strongest barrier, as illustrated by the distribution at (a) in Figure 6.9, then the solution of equations (6.6) and (6.7) under the assumptions postulated above is effectively determinate. At first sight it would appear that modern theory allows no variability! Given the strain level, $\Delta\gamma_p$, all members of the population have the same life; or, expressed inversely, at a given life, N, all members of the population have the same fatigue damage resistance measured as the plastic shear strain range $\Delta\gamma_p$. This quantity is apparently used in the propagation equations because it is related to the crack-growth direction. It is not, however, what designers use. The fatigue resistance is always expressed for design purposes in terms of the **stress range**, Δs, because it is the load that is usually known and the size of the part that has to be determined. Of course, strain can be converted into stress, or vice versa, using the cyclic stress–strain behaviour of the material, including the cyclic yield.

The cyclic yield differs from the static yield by virtue of the strain-hardening that takes place during early cycling: see Manson (1966). It has already been shown (see section 3.3) that strain-hardening does not vary much from one item to another in a given material population. It follows that, although the mean cyclic yield may differ from the mean static yield, its variation about the mean, i.e. the standard deviation of the cyclic yield, must be roughly equal to that of the static yield. Cyclic yield is therefore certainly a distributed quantity. The same plastic shear strain would thus arise from different stresses in different members of a population. It seems that modern theory is suggesting that the variability in the fatigue resistance measured as a stress is mainly a consequence of the variability in the cyclic yield. Such a conclusion would not be in conflict with the earlier work of Coffin, Manson, and others who also correlated the fatigue resistance with strain. If correct, this provides an important lead: for if the variation in stress s arises mainly from the variation in cyclic yield, then the variation of s in Figure 6.7 is the same for all values of N, and thus equal to the variation of the fatigue limit. The standard deviation of the fatigue resistance is thus constant for all values of N, and equal to the standard deviation of the fatigue limit. Furthermore, because the variation of the cycle yield is substantially the same as that of the static yield, the variation of the fatigue limit must also be equal to the variation of the static yield and hence the variation in the ultimate tensile strength. Materials science leads to the surprising conclusion that the variation about the mean, and hence standard deviation, of the fatigue resistance (damage resistance) at all values of N including the limit, of the yield strength, and of the ultimate tensile strength are, at least to a first approximation, all the same! Furthermore, there is a one-on-one ordered relationship between the individual items of all these distributions – provided, of course, that the incipient crack lengths are small.

What if the incipient cracks are not small? Consider the incipient crack distribution shown at (b) in Figure 6.9, in which some crack lengths are greater than d. In all circumstances for which $a_0 > a_*$ the long dormant state shown in Figure 6.9 will be omitted completely: the crack forms immediately as a physically small one following the growth pattern modelled by equation (6.7). The lives of all those items for which $a_0 > a_*$ will thus be considerably less than those of the main population: see for example the item from distribution (b) represented by the coarse dotted line in Figure 6.9. Reference to this figure shows that only a very small change in initial crack length is required to move a particular item from a long-life regime to a short-life regime, giving rise to a virtual discontinuity in life. There are, in effect, two populations, or subsets of the initial population, separated by an initial crack length roughly equal to d. This phenomenon was first recognised experimentally by Bompas-Smith (1969), who used the term **knee** to describe the manner in which such

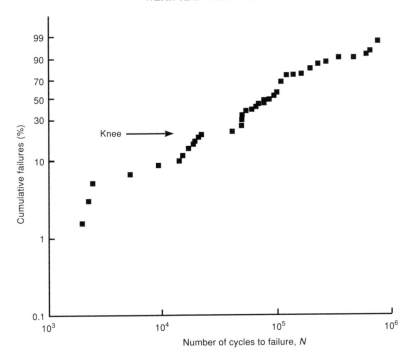

Figure 6.10 Fatigue tests at constant stress on strips from aero-engine compressor discs. Material: maraging 12% Cr steel. Data from Bompas-Smith (1969).

data plotted on Weibull scales. A typical Bompas-Smith curve is reproduced in Figure 6.10. It shows fatigue test data on some gas turbine compressor disc material. The general shape of this test curve is not unusual. It has the effect of reducing quite considerably the life of the weaker items of a population, i.e. the N_F life for small values of F.

Bompas-Smith was able to identify those items that made up the knee by a non-destructive (magnetic) test. Figure 6.11 shows the life distribution of the same material with the defective items removed. The knee has gone. It confirms the association of the knee with 'large' initial defects. Some workers use the expression **roller-coaster** rather than 'knee' to describe the consequences of large initial defects. The description stems from the undulating nature of the resulting hazard curve – see Figure 6.12. The phenomenon is all too often apparent in field data on components: see for example the data (Carter, 1986) on vehicle propellor shaft universal joints plotted in Figure 6.13. In addition to the obvious discontinuity between the two distinctive parts of the plot, the lower limb has a very characteristic reversed curvature, which does not conform to a distribution with a negative locating constant when plotted on Weibull paper. (The reason for this shape is discussed in an

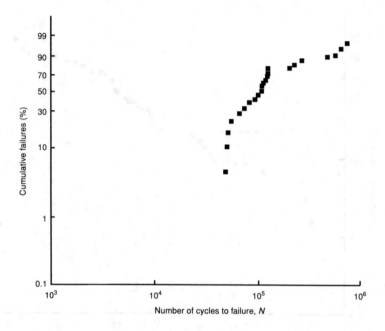

Figure 6.11 Fatigue tests using same material as in Figure 6.10 with pieces free from magnetic indication. Data from Bompas-Smith (1969).

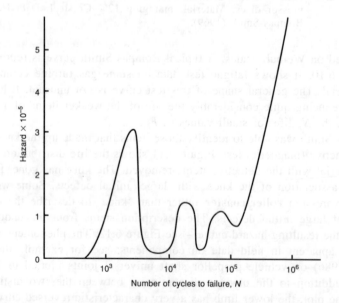

Figure 6.12 A roller-coaster based roughly on data from Figure 6.10. Troughs on roller-coaster correspond to plateaux on curves of Figure 6.10.

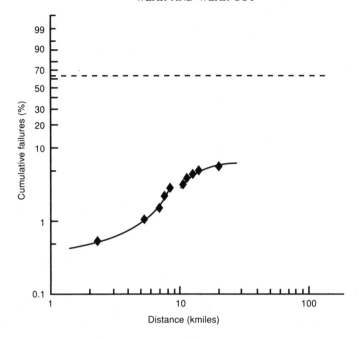

Figure 6.13 In-service fatigue failure pattern of propeller shaft universal joints. Total number of vehicles in survey = 180.

appendix to this chapter.) For the complete population the initial crack length dominates the failure distribution. It is impossible to calculate that part of the whole life distribution that forms the knee without a knowledge of this initial crack length distribution. Such information is certainly not available at present. The major consequence of the behaviour is that a constant plastic shear strain cannot be associated with a constant life. The damage resistance at constant N depends significantly on other factors besides the cyclic yield, and hence its distribution (and its standard deviation) cannot be constant for all values of N. The simple conclusion reached at the end of the previous paragraph is invalid, and applies only to a product in which every item satisfies the condition that $a_0 \ll d$. Such a product will henceforth be described as **well-finished homogeneous**.

Similar behaviour can arise when accidental damage is done to an already well-finished product, as for example when being assembled into a parent machine. Scratches or similar damage of length greater than d can be inflicted. Such damage will result in the behaviour just described, and the damaged items constitute a markedly low-life subset of the product. Clearly, knees or roller-coasters cannot be tolerated. This is really a matter for quality control rather than design, though it will be further examined under design in Chapter 7. Meanwhile we return to the well-finished homogeneous product.

The particular need is for test evidence to substantiate the constancy of the damage resistance (stress) variation at all values of N, for the reader must be well aware that some sweeping assumptions have had to be made in reaching this conclusion. Unfortunately, I have not been able to find sufficiently detailed test evidence to confirm or refute the conclusion convincingly. Of the data available, most are statistically biased inasmuch as experimenters have conducted different numbers of tests at different stress levels. This is almost a necessity if one is trying to determine the 'limit' by means of tests at constant stress using a practical number of testpieces. Other data show definite signs of knees or roller-coaster characteristics and must be discarded. Many test results have too few data points to justify any analysis.

Perhaps the best set of data is that published by Cicci (1964). The original data are reproduced in Figure 6.14. Some 50 data points (48, 44, 47, 48) were obtained at each of five stress levels on pieces cut from a single sheet of maraging steel. The data cannot thus be regarded as representative of the material in general, but at least the source of the testpieces implies a single population, and the number of tests adequately defines the central distribution at each stress level. Analysis of the data shows no 'knees'. (There is a faint suggestion of a knee at 110 klbf/in^2, but this is not reflected at any other stress level.) On the debit side the number of stress levels explored is restricted, and it is impossible to define the distribution of any damage threshold from these data. However, it is possible to evaluate the distribution of the damage resistance (stress) at constant N over an acceptable range of N values with reasonable confidence.

A reinterpretation of the data in this form is given in Figure 6.15. For this purpose the data at each constant stress level were ranked, using median ranks, in the usual way. No smoothed curve or parametric distribution was fitted to this estimate, however, but the rank corresponding to specified s and N values was obtained by linear interpolation between the two data points most adjacent to those s and N values. In this way scatter was not smoothed out, and is carried forward. The values obtained are plotted in Figure 6.15. The points plotted in Figure 6.15 thus give the cumulative probability of failure at different values of N using stress as the variate: the curves plotted in Figure 6.15 are all obtained by translation of a common curve along the stress axis: i.e. the mean stress is changed but the variation about the mean is kept constant. Apart from one point, all lie on, or reasonably close to, the curve translated for the appropriate N. It is therefore concluded that the variation of the fatigue resistance remains the same, as suggested by the theory, for all values of N within the range tested.

It is impossible to say, of course, whether the same distribution would apply to the fatigue limit if the testing had been extended to lower stresses on the one hand, or to the ultimate tensile strength if testing had been extended to higher stresses on the other. However, there is no evidence of any systematic trend with N in the data of Figure 6.15. I am happy to accept that the

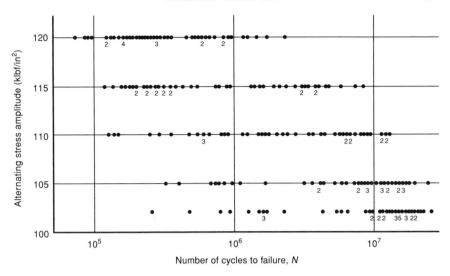

Figure 6.14 Rotating bending fatigue test results on air-melted, vacuum-degassed maraging steel. All specimens from single sheet. Data from Cicci (1964).

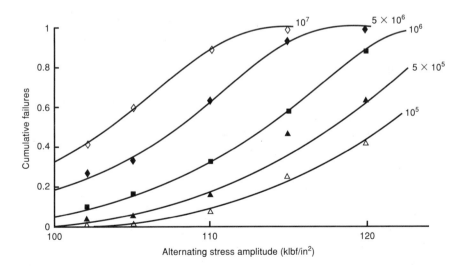

Figure 6.15 Replot of data in Figure 6.14 showing constancy of stress distribution at all N in range tested.

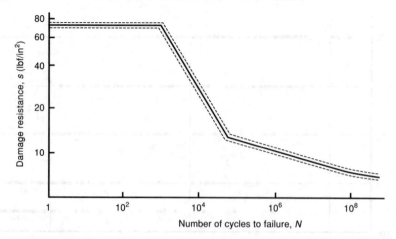

Figure 6.16 Bore's scatter band for the fatigue damage resistance of a DTD 364 component. Stress concentration relative to polished bar = 4. Data from Bore (1956).

constancy of the variation is maintained to higher values of N, and hence possibly to the 'limit', because the propagation equations hold over this extended range. I am much less certain that extrapolation to lower N is valid. Any neglected variation in the propagation equation parameters, resulting in a significant variation in the plastic shear strain, would give rise to an additional variation in the damage resistance. The additional variation may not be constant for all N, of course, but would apply to all $N > 1$. It would not, however, apply to the ultimate tensile strength, which is not modelled by the propagation equations. The equality of the variations in the UTS and the damage resistance for all values of N (except unity) is therefore a crucial test of the theoretical approach, which is not addressed by Cicci's test results.

Although the above is probably the best of extremely limited evidence, the constancy of the s distribution was noted as long ago as 1956 by Bore, who presented evidence on a DTD 364 component 'showing relative constancy of stress scatter, compared with endurance scatter'. (For 'scatter' read 'range'. Bore seems to have eschewed a statistical interpretation: he 'supposes' that his scatter band excludes about 1 in 100 components, i.e. lies somewhere in the region of $\pm 2\sigma$ from the mean.) He claims that this constancy extends to $N = 1$, i.e. applies to the ultimate tensile strength. He presents only linearised curves – see Figure 6.16 – without raw data points, so it is not possible to reinterpret his data statistically, and one does not know what weight to place on his observations. These were nevertheless made in a different context from the thesis being advanced in this book, and are therefore totally unbiased in the current context. Furthermore, Bore was the first, so far as I am aware, to describe the extended life distribution, which is discussed later. So his views on distributed fatigue behaviour cannot be dismissed lightly.

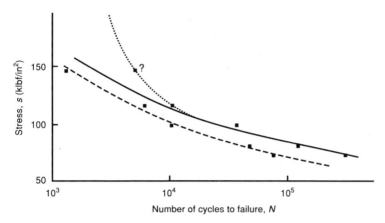

Figure 6.17   Reversed bending fatigue tests on SAE 4340; $R(c) = 35/40$. Number of tests at each stress level $= 18$ except at level marked? $= 12$: ——, mean plus 3 standard deviations (ignoring queried point); ⋯⋯, mean plus 3 standard deviations (including queried point); ---, mean minus 3 standard deviations. Data from Kecicioglu (1983).

Apart from Bore's remarks, there is some colloquial evidence to suggest that the variations of the fatigue resistance and the ultimate tensile strength are the same. Thus one commonly hears statements that the coefficient of variation of the fatigue limit can be expected to be about twice that of the ultimate tensile strength (whether this means '×2' or just '>' is open to conjecture). It is much more firmly established that the fatigue limit of steels is about half the ultimate tensile strength. If both these statements are literally correct, then the standard deviations of the two quantities must be equal. The reader must treat such anecdotal evidence for what he or she thinks it is worth.

The only properly recorded evidence in this connection was reported by Kecicioglu (1972), who presented the results of fatigue tests in reversed bending on notched specimens of SAE 4340 $R(c)$ 35/40. Like Cicci, he gave results of tests at five different stress levels, but only 18 pieces were tested at the lower four levels and 12 at the highest. A log-Normal distribution was fitted to the data at each stress level: note that this becomes the ubiquitous Normal distribution when life is measured as log N. His analysis was rigorous by classical statistical standards, but the reader is reminded that over 1000 data points are necessary to define the $\pm 3\sigma$ points with any confidence: 18/12 is certainly not enough. No reference was made to knees or roller-coaster characteristics. In view of the rigour of the analysis one may reasonably conclude that none was observed, but in view of the number of tests that does not mean they were non-existent.

From the constant stress distributions lives at $\pm 3\sigma$ from the mean were estimated. Kecicioglu's reported values are plotted in Figure 6.17 using semi-logarithmic axes. It is, however, difficult to draw curves through the points in Figure 6.17 because of a doubtful point (marked '?' on the figure). A full line has been drawn ignoring this point, and a dotted line assuming it to be valid. The latter leads to absurd values of the damage resistance at 1000 cycles and so is momentarily ignored. More tests at the highest stress level are required, or better still tests at an even higher stress level to confirm the trend, whatever it may be. The $\pm 3\sigma$ lines in Figure 6.17 are separated by a constant stress of 13 750 lbf/in², representing 6σ, over the full N range. So the standard deviation of the damage resistance is 2292 lbf/in², and is constant from, say, 2×10^3 to 5.5×10^5 cycles. The constancy of the standard deviation is in agreement with Cicci's work.

However, Kecicioglu carried it a stage further, and measured the distribution of the UTS of SAE 4340 $R(c)$ 35/40, though this time using even fewer testpieces: ten. He quoted the standard deviation of the UTS as 2500 lbf/in². This is about as good an agreement with the standard deviation of the damage resistance as one could expect. Statistically, the difference between the standard deviations is not significant. Furthermore, the standard deviation of the UTS is greater than that of the damage resistance: not less, as one would expect if the theory were in error. It would thus appear on the basis of this very limited evidence that treating the propagation equation parameters as more or less deterministic is fully justified, and the standard deviation of the damage resistance for all N is equal to that of the UTS. It may well be objected that this conclusion is biased, in that a test point has been ignored for no objective reason. If that test point is valid there is a significant difference between the two standard deviations at the 5 per cent confidence level, for N around 5×10^3. The difference is not significant at the 2.5 per cent level. Taking Kecicioglu's data at their face value, a strictly statistical conclusion is that we should suspect that the theory is in error, though we cannot really be sure. Even so, it could be claimed that this 'unbiased' statistical conclusion is itself based on biased data, in that an essential test series at a higher stress level was not carried out. It is thus extremely doubtful whether the theory could be rejected on this evidence, and the previous conclusion could well be the stronger interpretation.

To sum up: the theoretical conclusion that the variation of the damage resistance about its mean is independent of N is substantially confirmed by test evidence, at least over a wide N range. The confirmation is less certain at low values of N, particularly at $N = 1$. Statistical appraisal is indeterminate at $N = 1$, but an engineering judgement would be that acceptable test evidence agrees with theoretical predictions.

If that can be accepted, it is possible to construct the complete s–N distribution for a well-finished homogeneous material, given the standard s–N median curve derived by standard contemporary practice from test and the variation of the ultimate tensile strength, which is relatively easily measured at the F levels usually required. Figure 6.18 shows how any s–N_F curve is related to

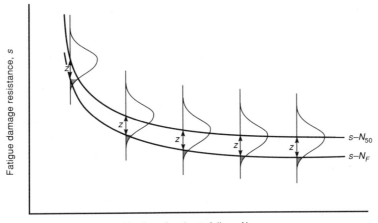

Figure 6.18   Derivation of s–N_F curve from standard s–N (s–N_{50}) data. $z = s_F - s_{50}$.
          z is constant for all values of N at a given value of F.

the standard median s–N_{50} curve. The theoretically constant stress decrement, z, equals the difference in the stress (variate) corresponding to the cumulative probability F and the median, i.e.

$$z = s_F - s_{50} \tag{6.11}$$

If, for example, the fatigue limit is Normally distributed, then z for an F equal to 10 per cent will be -1.28 standard deviations. In the case of a non-parametric representation of the distributions, z will have to be determined graphically or numerically. For a well-finished homogeneous population the function $\zeta(s)$ describing the whole set of s–N curves can then be written as

$$N = \zeta(s-z) \tag{6.12}$$

Putting $z = 0$ allows the function $\zeta(s)$ to represent the standard median curve, and so it can be fitted to test data.

When test data are not available it is often taken as an inverse power relation:

$$N = \frac{\text{constant}}{s^m} \tag{6.13}$$

The index, m, is about unity for simple erosive wear, is said to be about 3 for bearings, 4 for roads, 5 for gears, and can range up to 9 or 10 for general fatigue. Care needs to be taken with these figures, which may apply only to that part of the s–N curve relevant to the application. Alternatively, equation (6.13) can be fitted to the fatigue limit at 10^6 (or 10^7) cycles, and to the ultimate tensile strength at 10^3 cycles. Manson (1966) suggests that a value of 0.8 times the ultimate tensile strength is more appropriate at 10^3 cycles. Values of the constant and m in (6.13) are defined by the two points. Another alternative is a linear curve on semi-logarithmic axes, giving

$$\log N = c_1 + c_2 s \qquad (6.14)$$

where the constants c_1 and c_2 are determined by fitting to the end points of the s–N curve as described above.

As a consequence of the variation of E about its mean being the same at all N, and the one-on-one ordered relationship between all the items of the relevant distributions, the ranking of all items of a well-finished homogeneous population is the same at all N. Thus Miner's rule can be applied to any s–N_F curve in these circumstances, confirming the assumption made earlier, albeit within the restrictions of a well-finished homogeneous population.

It is suggested that in spite of only embryo theory and little evidence, a sufficiently good base has been established upon which to build a description of the wear process, at least for fatigue.

It is apposite to record here that the characteristics of the fatigue damage resistance recorded above differ significantly from that generally assumed. The latter has been illustrated by Figure 6.5, which conforms with much of the literature – see for example Nixon (1971) as a very authoritative source – and draws attention to the supposed high scatter (standard deviation) of the fatigue limit compared with that of the ultimate tensile strength. Figure 6.5 should be compared with Figure 6.16 (Bore's data) and Figure 6.18 illustrating the outcome of the theory presented earlier in this chapter. The standard deviation is constant, and does not behave as illustrated in Figure 6.5. It is this property that will enable us to develop a fairly simple statistical design methodology for the wear processes.

Before leaving the s–N curve, it is important to note that the quantity s plotted on all s–N curves is the stress that, when kept constant, will result in failure after N cycles. It is *not* the strength at any instant: for example, when the number of applied cycles, n, is less than the number of cycles, N, required to cause failure, the part is then still intact. It is not therefore the stress used in load–strength interference modelling. For that purpose one must use the instantaneous strength, i.e. the ability of any item to withstand a single further application of load, recognising that on account of the damage done by previous applications of load, the strength may be less than its initial value. It is often referred to in fatigue literature as the **residual strength**. In this book the word 'strength' has always been used with the above meaning, which is the only one that makes any sense in reliability theory – see Figure 6.2. The value plotted on the s–N curve has consistently been called the **damage resistance**, even though one finds it called the fatigue strength both colloquially and in many formal texts. Figure 6.2 clearly shows that strength and the damage resistance are not the same

According to modern theory, the only change in the state of the material during the fatigue process is the presence of the crack or cracks, though as far as I can see there could be a lot of them. Certainly, the microstructurally

short crack is so small that it cannot detract from the strength (i.e. the ultimate tensile strength) of the item. Because the physically small crack is initially of the same length, it too cannot immediately detract from the strength,
which will retain its initial value until that crack has grown sufficiently to
reduce the cross-sectional area significantly, or until the fracture toughness is
exceeded. Reference to Figure 6.9 shows that this can occur only in the latter
stages of crack development. It follows therefore that, so far as load–strength
modelling is concerned, the item will retain its initial strength until the rapid
crack growth towards the end of life brings about a collapse of strength over
a relatively small number of load applications. Consequently, the strength distribution remains unchanged over the failure-free period.

We can also assess the strength to be used in any load–strength interference modelling using Miner's rule. Let $L(s)$ be the distribution of stress (pdf),
measured at the critical section, arising from the application of load. Let the
stress causing failure in a further single application of load, i.e. the strength,
following applications of i individual load stresses of values L_i taken at random from the distribution $L(s)$, be S. Let N_i be the number of applications
required to produce a failure at L_i, and let N the number at S. Then Miner's
rule can be written as

$$\frac{1}{N} + \Sigma \frac{n_i}{N_i} = 1 \tag{6.15}$$

or

$$N = \frac{1}{1 - \Sigma(n_i/N_i)} \tag{6.16}$$

where n_i is the number of load applications at L_i. The instantaneous or residual or single load strength, S, i.e. the strength to be used in load–strength
interference modelling, is then the stress s corresponding to N on the s–N
curve.

It should be noted that, if $\Sigma(n_i/N_i)$ is as high as 50 per cent, then N is only
2, and hence the instantaneous strength at 50 per cent life is virtually equal
to the initial strength. Likewise, it follows that the strength at 90 per cent of
the life is that corresponding to ten applications of stress, and even at 99 per
cent of life is that corresponding to only 100 applications. At 99.9 per
cent of life it is that corresponding to 1000 applications. Thus the instantaneous strength remains almost unchanged throughout virtually the whole life.
This is illustrated in Figure 6.19(b), which is based on the fatigue s–N curve
shown at (a), and has been deduced using equation (6.16). However, at 100
per cent of the life, when $\Sigma(n_i/N_i) = 1$, the instantaneous strength is that at an
infinite number of cycles, i.e. the threshold value. If this is zero then the
strength is zero. The collapse of strength takes place over very few cycles, as
demonstrated in Figure 6.19(b). As a close approximation, the strength can be
assumed to be given by the dotted line in that figure, i.e. constant for all load
applications followed by an instantaneous drop to zero. This behaviour is

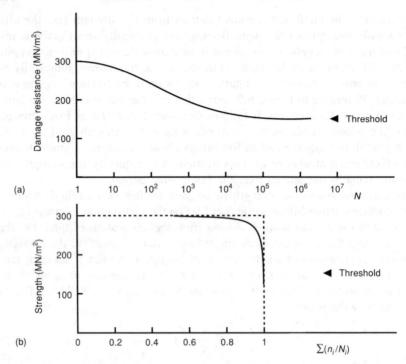

Figure 6.19 Damage resistance s–N curve and corresponding instantaneous (single load) strength.

sometimes described as **sudden death** in the literature. Even if the damage threshold is not zero and the strength at $\Sigma(n_i/N_i) = 1$ is finite, failure must very soon follow, because loads greater than this must have been applied to cause damage in the first place. If the load is constant the next application must essentially cause failure. Sudden death is to be expected in all cases.

How far is all this reflected in the actual behaviour of materials? Fortunately, Talreja (1979) has investigated this experimentally in some detail, at least for one material. He subjected artificially flawed testpieces of a Cr–Mo–V steel to various amounts of random loading and measured the residual strength. His results are reproduced in Figure 6.20. The first graph gives the initial undamaged strength (the ultimate tensile strength), and the other four graphs give the residual strength after increasing amounts of damage. The dotted line on each of these four graphs is a replot of the initial strength for direct comparison (with the upper branch of the data). The abscissa used by Talreja for these graphs is given by

$$Mz_i = [-\ln(1 - MP_i)]^{1/\beta} \tag{6.16}$$

where MP is the median probability, i.e. the probability based on median ranks, and β is the Weibull shaping parameter, so that straight lines on these

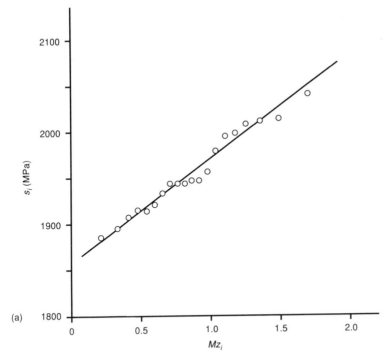

Figure 6.20 Instantaneous (residual) strength of Cr–Mo–V steel after various amounts of fatigue damage: ———, curve after damage; ------, no damage. (a) Initial strength; (b) residual strength after 50 periods; (c) residual strength after 100 periods; (d) residual strength after 150 periods; (e) residual strength after 200 periods. Data from Talreja (1979).

axes correspond to a Weibull distribution. As the damage is increased, the number of test points decreases because the test pieces failed by fatigue during the damage process (two failed at 150 periods of fatigue damage and 11 at 200 periods).

Talreja attributed the discontinuity at an Mz_i of about unity after 100 periods of damage to a similar mechanism that is responsible for the well-known delay in crack growth after overloading: the high loads in the distribution acting as 'overloads'. While this is a feasible explanation, I note a corresponding, though less intense, discontinuity in the undamaged strength. It is repeated to some extent in the other plots. In spite of these minor perturbations, a comparison of the full and dotted lines in Figure 6.20(b)–(e) indicates a close agreement between the initial and instantaneous or residual strength distribution for all degrees of damage, as suggested by theory. Certainly, the instantaneous strength is never less than its initial value. The slope of the lower branch on each of the four graphs indicates that the strength collapse is not quite instantaneous, but even so it is a pretty quick affair. It

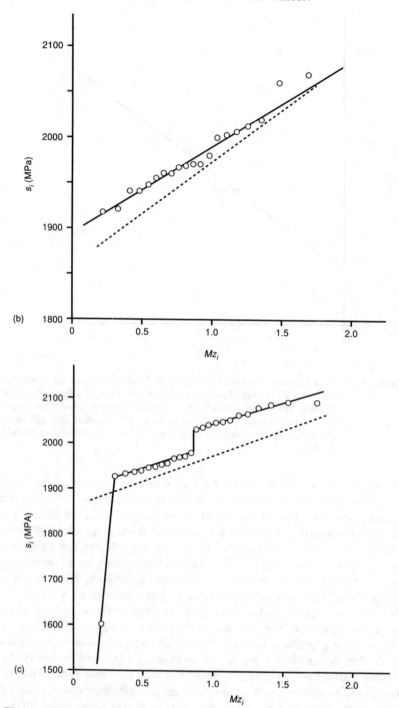

Figure 6.20   (continued)

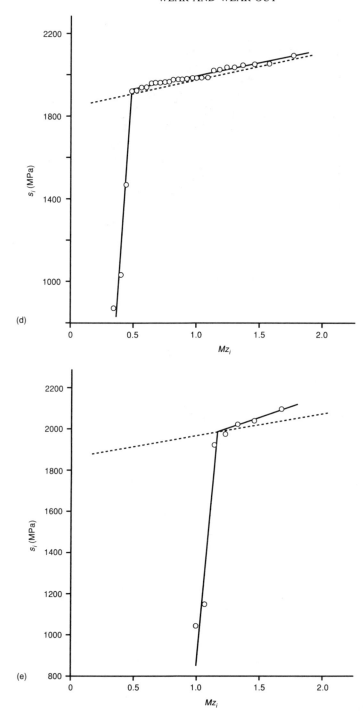

Figure 6.20 (continued)

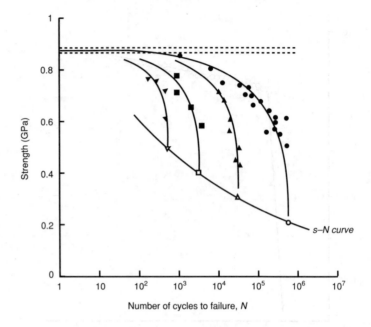

Figure 6.21 Instantaneous (residual) strength GRP 0/90 65% RH after tensile
fatigue cycling: ●, maximum stress 200MPa; ▲, maximum stress
300 MPa; ■, maximum stress 400 MPa; ▼, maximum stress 500
MPa. Data from Adam *et al.* (1986)

would seem reasonable to take it as instantaneous for design. It may safely
be concluded that for design purposes this material is behaving in accordance
with theoretical predictions.

Whether this is true of all materials is another matter. There are deterministic curves showing the same behaviour for many metals. Most engineers
have for generations assumed this behaviour to be self-evident. On the basis
of this evidence there is no reason to assume that Talreja's results are unrep-
resentative. On the other hand, Figure 6.21 shows some residual strength curves
for some composites. Although these are deterministic curves, they show a
more gradual loss of strength. There could be serious errors if a sudden death
behaviour were assumed for these materials. However, the failure of composites is probably better modelled as a series of discrete stress-rupture failures of the individual laminae making up the laminate, rather than as a continuous
process. Some care is obviously necessary before the theoretical model is
used universally. Nevertheless, for the modelling of most metals it is considered that the strength can be treated as constant, and the duration of the strength
collapse can be ignored: i.e. the eventual collapse is taken as instantaneous.
In that case the strength to be used in load–strength modelling is given by
the dotted line in Figure 6.19(b).

6.2.2 Other failure mechanisms

Considerable attention has been paid to the s–N distribution for fatigue, partly because it is often claimed to be the most important failure mechanism, and partly because consequentially it has been the most intensively studied. In this section the treatment will be carried forward to deal briefly with three other frequently recurring wear failure mechanisms: creep, erosive wear, and corrosion.

(a) Creep

There are two mechanisms of creep: **diffusion creep** (which gives rise to linear viscous creep) and **dislocation creep** (which gives rise to a power law). In spite of the above classification, diffusion plays an important role in both mechanisms.

Evans and Wilshire (1993) give a good introduction to creep, but do not discuss variation. Dislocation creep is the operative mechanism for most engineering materials at high temperatures, and diffusion creep is usually ignored in mechanical design. It will not therefore be pursued further here. In dislocation creep, plastic strain initially occurs, when a part is subject to continuous stress, by glide of dislocations along slip planes until some 'barrier' is reached. However, dislocations cannot be arrested indefinitely at the barriers, and at high stress and temperature the dislocation will 'climb' out of the slip plane, mainly by bulk diffusion of the atoms through the crystal. After climbing some distance, the dislocation may be able to proceed on a new slip plane until it meets another barrier, and so on. A new dislocation can now develop on the original slip plane, and so the creep continues. The rate of creep is largely determined by the rate of the diffusion process that allows the climb to take place. It is often modelled by the equation

$$\dot{\varepsilon} = A_c s^m e^{-(Q_c/R\tau)} \tag{6.18}$$

where $\dot{\varepsilon} = d\varepsilon/dt$ is the strain rate, A_c is an empirical constant, Q_c is the activation energy for creep = activation energy for lattice self-diffusion (Q_{SD}), R is the universal gas constant and τ is temperature. This creep process is often seen as a combination of work-hardening due to the deformation by slip under stress and a relaxation or softening due to the effects of temperature.

Now consider the position at the initiation of creep, which, like yield, starts by slip on glide planes. The material resistance to both processes is the same, because it arises from the same friction between the surfaces of the glide planes. So it is necessary for the diffusion process to reduce the frictional forces from the value associated with the yield to the value associated with the material's damage resistance, before creep can take place. But the diffusion process is controlled by Arrhenius's law (the exponential term in equation (6.18), in which the parameters are all basic physical constants and hence

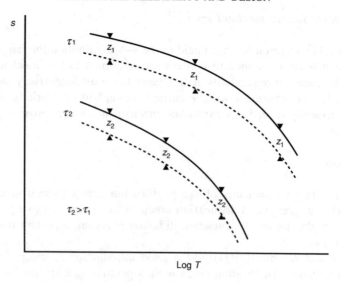

Figure 6.22 A schematic creep–rupture diagram: ——, standard curve for $F = 50\%$; ---, curve for $F\%$ fail. $z = s_F - s_{50}$.

deterministic. Thus the diffusion process is deterministic, and hence the variation in the initial creep damage resistance must be the same as the variation in the yield strength. Furthermore, because both this diffusion softening and work-hardening are more or less deterministic, the variation must remain the same throughout the whole life, or at least until the tertiary stage, when cracks may be formed and crack propagation could become the controlling process.

One thus concludes that, as for fatigue, the variation about the mean of the creep damage resistance is constant for all t (time under stress rather than the number of stress applications is the controlling parameter in this case), and is equal to the variation in the yield strength and hence that of the UTS. Creep-rupture or creep-failure diagrams can thus be easily constructed, as shown in Figure 6.22.

The above description of the stochastic nature of creep variability should be treated as speculative. There are no sufficiently detailed data available to me to either confirm or refute the treatment. It is interesting to note in this connection that, although Evans and Wilshire find it necessary to describe the least-squares method of obtaining regression lines from scattered data (i.e. acknowledging the existence of scatter), nowhere do they discuss the reasons for or the nature of scatter. Nowhere do they present a distribution of the properties involved or even quote a standard deviation. Scatter is measured as range! This is not intended as a criticism of their work – it is a true reflection of the state of the art. But it *is* intended as a criticism of the state of the art. The specialists apparently consider variation unimportant! Thus, with no supporting data, any theoretical hypothesising should be regarded only as guidelines in data collection and evaluation. *In extremis* theory could be used to calculate variations in the *s–T* curves for creep in exactly the same way as already described for fatigue. The reader must assess the risk of doing so for him or herself, knowing the particular circumstances of any application. There we must leave it.

(b) Erosive wear

Four mechanisms for erosive wear are usually quoted:

1. adhesive wear;
2. abrasive wear;
3. corrosive wear;
4. surface fatigue.

Any actual erosive wear situation probably involves a combination of all four mechanisms to a greater or lesser extent. Additionally, wear is often divided into rolling and sliding wear. Another description is fretting, which is small oscillating sliding wear. All these probably differ only as to the extent to which the above four failure mechanisms are individually involved.

Of the four mechanisms, corrosive wear is best dealt with under corrosion below, and surface fatigue has already been covered under general fatigue, leaving adhesive and abrasive wear for examination. Adhesive wear takes place by yield in shear of the surface asperities that have become locally 'welded' under contact pressure. The wear is therefore determined by the shear yield strength, and hence the wear resistance distribution is determined by the distribution of the yield strength. The standard deviation of the adhesive damage resistance would thus be constant throughout its life. Abrasive wear is more difficult to assess. It arises from the gouging action of hard particles broken off as a result of adhesive wear, and by particles introduced through the lubricant and other sources. Inasmuch as abrasive wear is a function of the hardness of the surfaces in contact, it too may be thought to have a constant standard deviation throughout the part's life. However, for both these wear processes the surfaces in contact are nearly always lubricated, and this is more likely to be the controlling factor. Even so, lubrication does break down, during start-up for example, and lubricants are impure. Additionally, the mating surfaces may have other roles to play, such as embedding the hard particles, and so on. The number of different situations is so large that it becomes impossible to generalise. However, we can be sure of one thing: sudden death describes the failure process – see Figure 6.23 for example. Numerous curves of this nature are available because monitoring of wear debris in the lubricant is a very common on-condition maintenance technique. Figure 6.23 shows that although the nature of the wear particles gives an indication of impending failure, break-up of the surfaces is very rapid indeed. Constant strength–sudden death, and therefore constant strength distribution throughout the product's

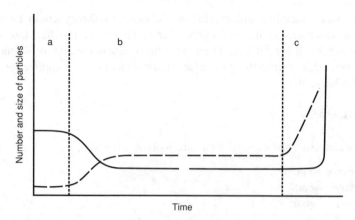

Figure 6.23 Erosive wear: (a) wear-in or run-in period; (b) constant wear during normal use; (c) surface disintegration leading to failure. ——, number of particles; ---, size of particles.

life, is a safe design assumption, but a confirmatory test of the assembly is the only sure way to obtain design data.

(c) Corrosion

The underlying physical process of corrosion has been well understood for a long time – far better than the mechanism of fatigue. But generalised data or even a knowledge of the factors controlling its rate of growth (the equivalent of equations (6.6) and (6.7)) are sadly lacking. There may be good reasons for this. In the first place there are very many more controlling variables than for fatigue, and covariation plays an important role. Even more testing would be needed than for fatigue – considerably more. Second, corrosion of itself is not often the cause of failure. It may lead to a weakened structure, it may lead to blocked pipes, it may lead to seizure, and so on: there are endless possibilities. The amount of unacceptable corrosion and the failure criterion can differ widely. Added to this, corrosion failures are often not due to the corrosion itself but to the failure of some prophylactic measures, which allows the corrosion to proceed. The very presence of corrosion may then be taken as failure. The actual failure mechanism in these circumstances can be mechanical not chemical. Finally, the results of corrosion are only too obvious. Corrosion may not have been observed before a failure, but nearly always was observable had the proper inspection processes been followed. Coupled with the very variable nature of the process, this all leads to an on-condition maintenance policy as the prime action for its containment. Current design is essentially deterministic: if two alternatives are available to control corrosion and the mean of one is better than the mean of the other then use the better, and let maintenance look after the variability. This has a certain amount of logic behind it, but it is not design for reliability, though some concessions to reliability are often made in the form of some worst-case scenario.

Current lack of data is possibly engendered by the economic belief that it is not worth collecting data (expensively) for a failure mode that can be more cheaply countered by simple inspection. This outlook undoubtedly influences the design approach to many of the wear processes whose progress can be evaluated by inspection. It is a tenable point of view, and the economics of achieving reliability will be examined more closely under design strategy. At present, it would seem that the only method of evaluating the reliability of any corrosion process is by test – of an actual part not a testpiece – in an actual environment.

The three examples illustrate in outline how theory can or cannot be used to obtain some indication of variability. For well-studied failure modes, such as fatigue, theory may allow tentative quantitative estimates, but for the less-studied modes theory is of little use. Distributions can always be determined by test, of course, but that is both expensive and time consuming. The advantage of a valid theory is that extrapolation of limited test data becomes possible. But theory has an additional role to play. Wong (1991) has shown that roller-coasters (knees) occur with many failure modes in addition to fatigue. A proper theoretical understanding of any wear process tells us whether such characteristics are possible and how to identify them. This is essential, for such behaviour can completely invalidate any design method that concentrates on the bulk property of the material.

Obviously, there are many wear mechanisms besides those already discussed, each requiring a different theoretical approach. Worse still, there are mixed modes. It is impossible to cover them all in one short book. Designers must clearly become conversant with the mechanisms that are relevant to their own product, and themselves develop the necessary theoretical and experimental support. So $s–N$ or $s–T$ distributions will not be pursued further here, but later a case study will be presented, after design background and objectives and hence requirements have been examined in more detail, to illustrate how theory can help to sort out inadequate test data.

Finally, it may be thought surprising that the distributions of so many failure mechanisms are related. The clue is that most of the failure mechanisms involving actual fracture start with the glide of a dislocation on a slip plane. The distribution of this phenomenon is the common link. Additionally, all follow a constant strength–sudden death pattern. This has been deduced using Miner's law, which can be applied to all mechanisms. Either this law must be in considerable error – errors by a factor of 10 make little difference up to about 90 per cent of the life – or the $s–N$ ($s–T$) curve must have an extraordinary shape in the early stages if sudden death is not to apply. There is a lot more common ground in the very many failure mechanisms than may be supposed at first sight.

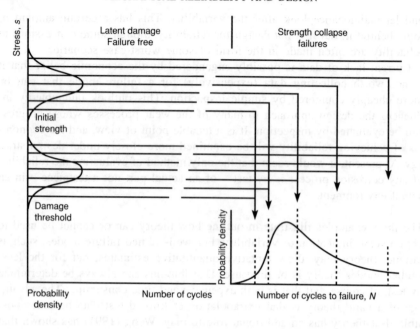

Figure 6.24 Diagrammatic representation of the wear process leading to failure.

6.3   The wear-out life distribution

A model of the wear process can now be formulated. It is presented diagrammatically in Figure 6.24. The curves plotted on the extreme left of this diagram show the distribution of the initial strength (which could be the ultimate tensile) and of the damage threshold.

It will be convenient to trace the history of seven sample items from the population. During the early applications of loads, damage will be done, but the strengths will remain constant as described in the previous section, i.e. the strength distribution remains unchanged. The strengths of the seven sample items will therefore be horizontal straight lines when plotted against age (cycles) as in Figure 6.24. Hence the load–strength interference will remain unchanged. If the population is initially intrinsically reliable, no measurable failures will occur. If the population is not intrinsically reliable, any failures that occur will be of a stress-rupture nature and not attributable to wear. It will be assumed that intrinsic reliability is achieved – as in the normal design situation illustrated at (a) in Figure 6.1. Then the regime described above will persist until the item with the lowest damage resistance reaches strength collapse and failure, as indicated in Figure 6.24. As more loads are applied, the remaining items of the population will fail in order of increasing damage resistance. It is interesting to note that the strength distribution at any arbitrary

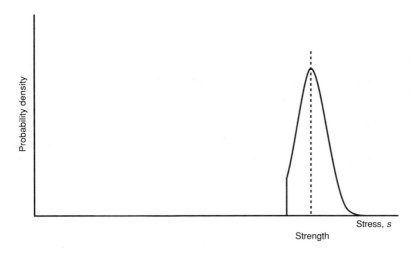

Figure 6.25 The strength distribution after about 10 per cent of the population has failed. Initial distribution as in Figure 6.2.

age will then coincide with the initial distribution less those (the weakest) that have failed. The strength distribution illustrated in Figure 6.2 becomes that shown in Figure 6.25 after about 10 percent failures, not that shown in Figure 6.1. The mean strength actually increases! As a consequence of the sudden-death behaviour, failure is a certainty, not a probability that has to be evaluated from load–strength interference. Load–strength interference model-ling is not therefore required, which considerably simplifies the subsequent treatment. Although individual lives to failure may be regarded as certainties, the different damage resistances of different items do give rise to a distributed life, as shown by the pdf at the bottom right of Figure 6.24. The probabilistic damage threshold distribution has been transformed into a probabilistic life distribution.

There is a one-on-one ordered relationship between the individual items making up the damage threshold and the life-to-failure distributions. To de-termine quantitatively the distribution of life this transformation must be evalu-ated: i.e. the value of N corresponding to each E has to be calculated.

For this purpose it will be assumed that it is valid to apply Miner's rule to each $s-N_F$ curve in order to estimate individual lives: i.e. in the case of fa-tigue the population is assumed to be well finished and homogeneous. Any other necessary condition must be satisfied for other wear processes. If this is not so, no simple solution is possible.

If the median curve is given by

$$N = \zeta(s) \tag{6.19}$$

it has been shown (equation (6.13)) that the set of generalised $s-N_F$ curves can be written as

$$N = \zeta(s - z) \tag{6.20}$$

where z is defined by equation (6.11) and in Figure 6.18. Consider an individual item drawn at random from a large population. Let its threshold resistance be E, where

$$E = \bar{E} + z \tag{6.21}$$

The damage done by the application of one load giving a stress of magnitude s_i is obtained from the damage law and equation (6.20) as

$$\Delta d_i = \frac{1}{\zeta(s_i - z)} \tag{6.22}$$

The number of times the load will be applied is

$$n_i = nL(s_i)ds \tag{6.23}$$

where n is the total number of load applications. Hence the damage inflicted by the particular load is given by

$$d_i = n_i \Delta d_i = nL(s_i)ds \; \frac{1}{\zeta(s_i - z)} \tag{6.24}$$

The total damage, d, inflicted by all the loads from the distribution $L(s)$ on an item of threshold damage resistance E is then the sum of the damage done by all the individual loads in the distribution $L(s)$, i.e.

$$d = \sum d_i = \int_0^\infty \frac{nL(s)}{\zeta(s - z)} \, ds \tag{6.25}$$

In equation (6.25) the suffix i applying to s has been dropped, and s takes all its values in the distribution $L(s)$.

By definition, $d = 1$ at failure. Therefore at failure

$$\sum d_i = \int_0^\infty \frac{NL(s)}{\zeta(s - z)} \, ds = 1 \tag{6.26}$$

where N is the number of cycles to failure of an item whose threshold strength is E. Whence

$$N = \frac{1}{\int_0^\infty \frac{L(s)}{\zeta(s - z)} \, ds} \tag{6.27}$$

Thus for each value of E given by equation (6.21) the life to failure, N, can be calculated.

It has already been noted that there is a one-on-one relationship between E and N. It follows that the number having a threshold damage resistance E will be the number failing at N, or mathematically

$$E(s)dE = f(N) \, dN \qquad (6.28)$$

where $f(N)$ is the probability density function of life to failure. Whence

$$f(N) = E(s) \frac{dE}{dN} \qquad (6.29)$$

This has been plotted on the bottom right-hand corner of Figure 6.24. The cumulative failures are given by

$$F(N) = \int_0^N f(N) \, dN \qquad (6.30)$$

and the hazard by

$$h(N) = \frac{f(N)}{1 - F(N)} \qquad (6.31)$$

Repeating for all values of E, the wear life pattern is completely resolved for any wear process, given the s–N distribution for the process.

It is recalled that load–strength interference modelling has not been used in the derivation of equation (6.27). This means that no statistical manipulation of the quantities involved has been used in arriving at that equation. The derivation follows standard practice, and equation (6.27) differs from more usual formulae only in that the variation in the quantities involved has been recognised and expressed in statistical terms. Thus if the load distribution is given in histogram form, then equation (6.27) becomes

$$N = \frac{\sum\limits_{i=1}^{i=I} n_i}{\sum\limits_{i=1}^{i=I} n_i / N_i} \qquad (6.32)$$

The reader is reminded, however, that the simple solutions of equations (6.24) and (6.32) are valid only if the stipulated conditions have been met.

Some life failure patterns evaluated by equations (6.27) and (6.31) are presented in Figure 6.26. Figure 6.26(a) shows a plot of the hazard against age for a linear s–N curve, i.e. $m = 1$ in equation (6.13). It depicts the ever-increasing hazard with age commonly associated with wear. Figure 6.26(b) shows the variation derived from a typical s–N curve ($m = 5$), which has zero damage threshold (no limit). Again the hazard increases, but much less rapidly, with age. Figure 6.26(c) shows the corresponding variation using the same s–N curve, but with a damage threshold introduced. This life failure pattern is somewhat different. After increasing from zero over the failure-free period, the hazard decreases slowly over a very long age span. It shows that wear should not always be associated with an increasing hazard, as is so often claimed in reliability texts. The number of variables, their behaviour,

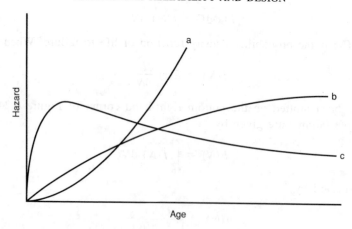

Figure 6.26 Some representative wear-out failure patterns: (a) $m = 1$ (equation (6.13)); (b) $m = 5$, no threshold; (c) $m = 5$, with threshold

and the number of combinations is so great that it is not possible to generalise on the shape of hazard curves. As a corollary, it is not possible to make any deductions regarding the wear process solely from the shape of the hazard curve.

The shape of the curve in Figure 6.26(c) can readily be accounted for physically. Suppose a non-distributed, i.e. constant, load is applied to a product having the s–N distribution in Figure 6.7. As the number of load applications is increased from zero there will at first be no failure, i.e. during the time that the microstructurally short crack is growing and is taken over by a physically small crack. From equation (3.36) the hazard is zero under these circumstances. Then any constant-load – constant-stress – line in Figure 6.7 will intersect $s–N_F$ curves of increasing F with increasing number of load applications, so that the cumulative failures will increase. From equation (3.36) this implies that the hazard must also increase from zero to a finite value.

Now consider the situation at very high numbers of load applications. The $s–N_F$ curves become straight lines of constant stress. Therefore no further intersections of the constant-load line with an $s–N_F$ curve can occur. Those items whose limiting value is less than the constant load will have already failed, and those items whose limiting value is greater than the constant load will never fail. The hazard will once again be zero – from equation (3.36) with $dF/dt = 0$.

Thus over the complete life the hazard will initially be zero but will then start to rise when failures occur. At some number of load applications it must reach a maximum, so that it can eventually return to zero at a high number of load applications: i.e. the shape of the complete curve corresponds to that in Figure 6.26(c). If the load is distributed, the final value will not of course be zero, but will approach zero as the loading roughness approaches zero.

In the next chapter the direct application of the above assessment of the wear process to design will be examined more closely.

References

Adam, T., Dickson, R.F., Jones, C.J., Reiter, H. and Harris, B. (1986) A power law fatigue damage model for fibre-reinforced plastic laminates. *Proceedings of the Institution of Mechanical Engineers*, **200** (C3), 158–166.

Bompas-Smith, J.H. (1969) The determination of distributions that describe the failures of mechanical components. *Eighth Annals of Reliability and Maintainability*, pp. 343–356.

Bore, C.L. (1956) The presentation of fatigue data for fatigue life calculations. *Journal of the Royal Aeronautical Society*, **60**, 321–346.

Carter, A.D.S. (1986) *Mechanical Reliability*, 2nd edn, Macmillan Press, Basingstoke.

Cicci, F. (1964) An investigation of the statistical distribution of constant amplitude endurances for a maraging steel, *UTIUS Tech Note No. 73*, UTIUS, New York.

Evans, R.W. and Wilshire, B. (1993) *Introduction to Creep*, The Institute of Metals, London.

Kecicioglu, D.B. (1972) Reliability analysis of mechanical components and systems. *Nuclear Engineering*, **19**, 250–290.

Manson, S.S. (1966) *Thermal Stress and Low-cycle Fatigue*, McGraw-Hill, New York.

Miller, K.J. (1986) Retrospective and prospective views of fatigue research, in *Conference on Fatigue of Engineering Materials and Structures*, Vol. 1, Institution of Mechanical Engineers, London, pp. 6–11.

Miller, K.J. (1991) Metal fatigue – past, current and future, Twenty Seventh John Player Lecture. *Proceedings of the Institution of Mechanical Engineers*, **205**, 1–14.

Nixon, F. (1971) *Managing to Achieve Quality and Reliability*, McGraw-Hill, Maidenhead.

Talreja, R. (1979) On fatigue reliability under random loads. *Engineering Fracture Mechanics*, **11**, 717–732.

Wong, K.L. (1991) The physical basis of roller coaster hazard rate curves for electronics. *Quality and Reliability Engineering International*, **7**, 489–495. Also contains references to two companion papers.

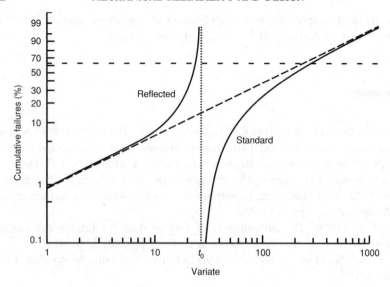

Figure 6.27   The relationship between a standard and a reflected
Weibull distribution.

Appendix: The Weibull representation of data below the knee

Data points below the knee all result from the propagation of a physically
small crack starting from a minimum crack length d (though more often greater)
and terminating at what may be taken as a unique crack length representing
failure. This implies that the amount of crack growth has an upper limit, and
hence the lives of the subset under consideration are bounded by an upper
limit. The limit is given by the number of cycles to grow from d to the length
for failure, because items having an initial crack length greater than d must
have shorter lives. Although the Weibull distribution has a lower bound at the
locating constant, it has no upper bound. It cannot therefore represent data
below the knee. The only possible distribution that could represent such data
plotted on Weibull scales would be a reversed Weibull distribution, i.e. a Weibull
distribution reflected about a locating constant. This would have an upper
bound, but no lower bound.

Figure 6.27 shows both a standard and a reflected Weibull distribution having the same Weibull parameters. The standard distribution is asymptotic to
the locating constant at low cumulative failures and to the two-parameter function
at high values of the variate. It will immediately be recognised as the form of
many Weibull raw data plots when truncated to the range usually covered by
those data. By contrast, the reflected distribution must be asymptotic to the
locating constant at high cumulative failures and to the two-parameter function
at low values of the variate. It is shown in Figure 6.27. It will be seen from

the figure that it corresponds closely to the shape of the curve below the knee in Figure 6.13 when truncated to the range of the data in that figure. The shape of the curve of data below the knee is therefore due to the unusual bounding of the data and its representation on Weibull scales. The unusual shape provides a useful indication of a data set having large initial defects.

7

Statistical design: wear modes

7.1 A simple design method

From the physical nature of wear behaviour developed in section 6.3, a very simple design technique can in principle be formulated for all wear processes. The life calculated using an $s–N_F$ curve in conjunction with any damage law is N_F. So, instead of using the standard median $s–N_{50}$ curve (50 per cent failures) to calculate N_{50} and relating that to N_F by empirical factors, the $s–N_F$ curve corresponding to the required $F\%$ failures can be derived as described in the previous chapter and used directly in design without any factor. In essence, statistically derived input data are being used in an otherwise conventional design methodology.

The procedure usually adopted is as follows. A distribution is fitted to the test data obtained at constant stress in a usual $s–N$ test. A Weibull distribution would be the first choice on account of its flexibility, but any other, such as the log–Normal, can be used if a better fit is obtained. As an illustration, five points at a constant stress level have been plotted on Weibull paper in Figure 7.1. These raw data are unlikely to lie on a straight line, and will have to be straightened, as in Figure 7.1, by one of the conventional techniques. This straight line can then be extrapolated to the specified failure level. For example, if a design of 95 per cent reliability is required, the life for $100 - 95 = 5$ per cent cumulative failures is read off the extrapolated straight line as in Figure 7.1. The process is repeated for the other stress levels, and an $s–N_5$ curve for 95 per cent reliability or 5 per cent failures (as contrasted with the 50 per cent for the standard $s–N$ curve) is produced. Such a curve is illustrated in Figure 7.2. This $s–N_5$ is then used directly for all design calculations relating to a 95 per cent reliability specification, without any factor of safety or reserve factor. The method has been illustrated graphically for clarity, but it would be more convenient to use one of the many computer programs available to fit Weibull distributions to the test data and then calculate the life for 5 per cent (N_5), 10 per cent (N_{10}), or any other cumulative failures from the three Weibull parameters. The method is fundamentally sound, and

124

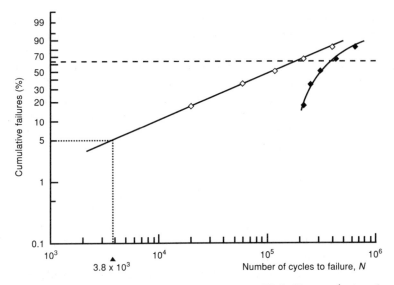

Figure 7.1   Plot of test data at constant stress on Weibull axes: ◆, raw data; ◇, straightened Weibull data. Weibull locating constant $= 0.2 \times 10^6$; $N_5 = 3.8 \times 10^3 + 0.2 \times 10^6 = 0.2038 \times 10^6$.

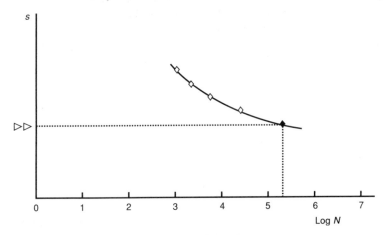

Figure 7.2 s–N_5 curve on semi-logarithmic axes: ◆, value of N from Figure 7.1; ▷▷, stress level of data in Figure 7.1; ◇, data at other stress levels.

is simple and straightforward – at least in principle. It is being explored by organisations in the UK and USA. In practice, the method may not be so simple or straightforward.

From the treatment of the wear process in the previous chapter, it will be appreciated that there are two conditions that must be satisfied for the method to be valid.

Condition 1 requires that the part should be intrinsically reliable, which in turn requires that the material should have a more or less constant strength while wear damage is being inflicted, and should conform to sudden-death behaviour. This ensures that no stress-rupture failures can arise. However, for designs aimed at a finite proportion of failures at life N_F, it is sufficient that intrinsic reliability be achieved using the strength distribution at N_F, i.e. the distribution described in Figure 6.25. This will ensure that none of the $(100 - F)$ per cent of the population that should fail by wear at a life greater than N_F could possibly fail by stress-rupture before N_F. The required condition is most likely to be met automatically for the vast majority of designs. An exception could occur in the case of very low-cycle wear in conjunction with a very low F to meet a safety requirement. Apart from this particular case, intrinsic reliability is in fact not a very stringent condition. Even if the strength does not remain truly constant during the wear process, it may still be met. It is essential, however, to check that intrinsic reliability is achieved at the specified design life in all cases.

Condition 2 requires that the material wear characteristics should have no knees or roller-coasters. This is a much more stringent condition. I am in fact alarmed at the high proportion of field data that show evidence of knees. Superficially, this implies that many design calculations would be invalid. However, a remedy is at hand. It has already been noted that knees effectively produce two populations, i.e. two subsets of the original population. Let the proportion of defective parts lying in the knee be F_d. Then the proportion $(1 - F_d)$ would have no defects and can be treated as a separate homogeneous subset of the total population, for which the above simple design methodology is valid. If R^* is the specified reliability

$$F^* = 1 - R^* \tag{7.1}$$

where F^* is made up of both unassessable defective failures in the defective subset and assessable failures in the homogeneous subset. If F is the proportion of assessable wear failures in the second subset, then

$$F^* = F_d + (1 - F_d) F \tag{7.2}$$

because $(1 - F_d)$ of the total population make up the non-defective subset, and all the defective failures must necessarily occur before any of the homogeneous failures. Hence the design target for the assumed homogeneous subset would now be F, which is given from (7.2) by

$$F = \frac{F^* - F_d}{1 - F_d} \tag{7.3}$$

The simple design method outlined above could then be adopted using this revised target reliability. In this way, non-homogeneous populations present no design obstacles, provided that F_d is known to be zero (achieved by rigorous quality control), or its value is known (achieved by extensive tests). Although this is an easy and valid method of dealing with the type of distribution under discussion, it should not be forgotten that in many instances in which rigorous statistical techniques are not adopted, the presence of knees is not known; F_d is, consequently, taken as zero. The neglected subset, F_d, can be the source of unacceptable failures in the field. F_d may well be the dominating features in many design situations. It must be known accurately before design can proceed.

There would appear to be three further difficulties, all associated with numerical accuracy, in the quantitative application of the technique:

1. As already discussed in Chapter 3, extrapolation of statistical distributions is not justified: i.e. Figure 7.1 may not be valid.
2. Sensitivity may preclude any reliance on the results.
3. Miner's rule may not be accurate.

Each of these is discussed separately and in detail below.

7.1.1 Distributions

The problems associated with the definition of a statistical distribution over the complete stress range have been covered in depth when dealing with stress-rupture mechanisms. The problems arising in the treatment of wear are no different, but fortunately there are some mitigating factors.

Wear failures are often countered by scheduled or on-condition maintenance, though some products may be allowed to run to failure before replacement (repair maintenance). If the first policy is followed it is often uneconomical to schedule replacement of all the items of a population before the first failure. This may not be true if safety is involved, and these circumstances have to be considered separately. A common optimum time for scheduled replacement is marked by the failure of about 10 per cent of the population. The actual figure will depend on many factors, but 10 per cent can be taken as a good first estimate. If the distribution of that first 10 per cent is not critical – the usual case – then the designer is concerned only with the upper 90 per cent of the population, i.e. with the population up to about 1.3 standard deviations from the mean. This is a far cry from the 4.75 or more standard deviations demanded by stress-rupture modes, and raises some hopes for numerical accuracy in wear mode design. Furthermore, any error in the achieved life pattern can often be countered by modification to the scheduled replacement interval. This may necessitate some departure from the ideal optimum, but reliability is under control. It is worth noting that maintenance policies are always optimised after a product has been in service for some time and its failure pattern is known from field data. No departure from current practice is thus required. For on-condition maintenance, it is clearly the main body of the population that dominates the activity, and some inaccuracy in the tail can be tolerated.

In these circumstances one might hope that the distributions could be defined, by test, over the whole practical range; so no extrapolation or use of parametric distributions would be required. To be more precise: if a designer wishes to be 95 per cent certain that he is using a measured minimum property that is achieved by 90 per cent of the population (i.e. only 10 per cent may have values less than the lowest measured value), then equation (3.27) shows that tests on 28 items are necessary – or it is possible to get away with 20 if one is prepared to be only 88 per cent certain that the minimum has been recorded. Even fewer tests would be required if a lower confidence were acceptable, of course. These are feasible figures for a practical test schedule, though still salutary. I have found that most engineers are prepared to accept, with some reluctance, that five tests at each stress level are necessary when defining a distributed s–N curve, but opposition increases rapidly as higher numbers are suggested. Even the lower number of 20 noted above would be regarded as research practice and not for routine test procedures. Furthermore, in some cases the above figures can be optimistic: for example, it is very difficult indeed to determine the distribution of the threshold, if that were required, by tests at constant stress. Rather more than the minimum can prove necessary in practice. Then knees can require considerably more tests if they are to be accurately defined. On the other hand, some short cuts are possible. For example, it is much quicker and cheaper to determine the variation of the damage resistance at low cycles to failure by testing at high stress, than it is at low stress and therefore high cycles to failure. Theory suggests that the same variation applies to all values of N, so it is not necessary to repeat at the time-consuming lower stresses. Pushed to the extreme, the theoretical hypothesis that the variation in the fatigue resistance is equal to the variation in the yield or ultimate tensile strength can be used. In that case only static tests on 20–30 pieces would be required to establish the variation. This is a very modest requirement. My own preference would follow this procedure but back it up by as many tests at one stress level as the organisation would accept, chosen as high as possible in the expected field range in order to minimise test time and cost. The objective of this procedure is twofold. First, it would ensure that all knees or roller-coaster characteristics were absent, or define their magnitude if present. Second, it can be used to check, using the condition that values estimated from the two distributions in Figure 6.5 must be the same, that the static distribution is valid – at least for one stress level.

Although there are difficulties, it does seem that, provided strict quality control is enforced so that knees and similar features are eliminated, and that the life of nothing much less than about 10 per cent of the population has to be defined, then the distribution of the damage resistance could be evaluated without too much testing, albeit somewhat more exacting than most practitioners would currently readily accept.

The essentially practical designer may well argue that insisting that all material properties are actually measured at the design probability is too exacting, and that some extrapolation of a parametric distribution fitted to the test data is justified. Indeed, it would be difficult to maintain that *no* extrapolation is possible in a design situation. But how much extrapolation? Clearly this will depend on the quality of the data. In the first place, it is necessary to ensure that any selected parametric distribution does fit the measured data. This must be achieved at the 50 per cent level in, for example, a Chi-squared or Kolmogorov–Smirnov goodness-of-fit test. The 50 per cent level is chosen to ensure that the critical value in either case is as likely as not to be exceeded: i.e. it is possible to attribute any difference between the value predicted by the selected parametric distribution and that observed on test to pure chance in random sampling. Failure to achieve this condition implies that some systematic variation is present, e.g. multiple suppliers, inadequate quality control, etc. Such data are of little use.

Assuming that the goodness-of-fit condition is satisfied, it is also necessary to be certain – confident – that the test data on which the fit is assessed do actually measure the true property distribution up to the initial extrapolation. (It is clearly possible to get a good fit to data that do not.) Thus a confidence level of at least 90 per cent should be demanded at that point. The required number of test points can then be calculated using equation (3.28). In these circumstances, I should have thought that extrapolation of about 1 standard deviation (but no more) is permissible for design purposes. The figure of 1 standard deviation is a subjective one, which cannot be evaluated by statistical techniques. Designers must make their own judgement of the appropriate value in each particular case, paying special attention to any extraneous factors that could affect the tail of the distribution.

How does this work in practice? If, for example, we are attempting to design for a 1 in 1000 chance of failure, i.e. at the three-sigma point for the Normal distribution, then the allowable 1 standard deviation extrapolation quoted in the previous paragraph requires that trustworthy data up to two sigma from the mean are provided. At this point, the probability of failure is 0.023 for the Normal distribution. Substituting in equation (2.38) shows that 100 test points are required at the 90 per cent confidence level. That number of test points is clearly not a practical possibility for, say, a fatigue test. However, if the theoretically postulated common variation of the UTS, yield, and damage resistance for some wear mechanisms can be accepted, then I would not regard a 100-point static test programme as outrageous if this level of reliability is essential. The distribution so determined would apply to all the relevant wear mechanisms. The number of data points can always be reduced by lowering the confidence level, of course: thus only 30 test are required at the 50 per cent (expected) level. But the designer needs to be sure that such measures are acceptable before embarking on that course. Still, it does seem possible to derive design data for up to a 1 in 1000 chance of failure by judicious extrapolation. For lower probabilities of failure, the required number of tests is likely to prove excessive.

It will be appreciated that extrapolation is possible only if it has already been established that there are no irregularities, e.g. knees, up to the maximum extrapolation point. This demands very rigorous quality control. Realistically, in many cases the statistical requirements will prove impossible to achieve; yet design must go on. Ridiculously low confidence levels may have to be accepted. Note: they are, effectively, in conventional design. Common sense must always prevail.

A case study of an actual situation will illustrate what can be achieved with limited data. At the same time, it will reveal some of the difficulties involved with real data, and some of the pitfalls that can invalidate a routine approach. The actual events took place some ten years ago, but will be interpreted as if contemporaneous with the writing of this book, i.e. using theory that was not available at the time.

An organisation with which the author was associated (British Coal) was concerned over the reliability of the gearboxes that formed part of their equipment, and suspected that empirical design methods used for the existing gears may have been at least partly responsible. In parallel with the techniques in current use, they proposed to use the method based on the relevant s–N_F curve for the design of future gears. Reliability would be specified quantitatively by the life for 10 per cent failures, i.e. by N_{10}. The majority of their gears were made from 832M13 (En36c in old nomenclature), and this material was selected for the first investigation. Because it was the properties of the as-finished material in the form of gears – not testpieces – that were required, it was proposed to test actual gears whose specification was representative of the range of gears in their boxes. This method of test would also provide an accurate representation of the load during both tooth engagement and disengagement. The gears would, of course, have to be designed by the organisation using their current technique. They would be manufactured by a gear specialist firm following the organisation's standard practice, which it intended to continue. The details of the test gear need not concern us here.

After much discussion it was decided that tests at five stress levels were necessary. This was a designer's requirement based on past experience. At each stress level six gears would be tested to failure in bending fatigue. This was far below a statistical requirement at any reasonable confidence level, but in view of the costs involved and the length of time for even such a limited test programme, this was all the organisation could afford. This seemed a reasonable compromise to me: reducing the stress levels to four, even if the designers could be persuaded to agree, would do little to improve the statistical confidence level if the same total number of tests were retained. It was proposed that a Weibull distribution, or some other if it proved more appropriate, should be fitted to the life (number of load applications on a tooth before failure) for each stress level, and extrapolated to give the life at 10 per cent cumulative failures as already described. The five stress levels would then provide five points defining the s–N_{10} curve, which would be used in design without any factor.

It is not possible to describe the test rig (a usual recirculating torque type), test technique etc. in this book for reasons of space, and the reader will have to accept my assurance that the results quoted below are a trustworthy evaluation of the gears tested.

We enter the case study at a point when testing has been held up far from completion. The position is this. The first batch of gears were received by the organisation, but the organisation's inspection rejected all but five of the gears. Their opinion was forcibly conveyed to the manufacturer! A second batch was subsequently received, but meanwhile testing went ahead at a failure-inducing bending stress level of 762 MPa on the first five gears passed as satisfactory. When the second batch was received, a further two tests at 762 MPa were completed and one suspended without failure at 10 million load applications. Another nine tests have been completed at 1272 MPa, all using second-batch gears. Testing has now stopped because there are no more gears, and the rig is engaged on other work. Typically, the testing is far behind schedule (due largely to the poor supply of acceptable gears), yet the s–N_{10} curve is required for design purposes. The gear life to failure recorded so far, expressed in millions of load applications per tooth and given in chronological order of test, is as follows.

Stress level 762 MPa:

0.677   10.83   0.533   2.30   0.642   6.5(P)   5.7(P)   10.0(S)

Stress level 1272 MPa:

0.23   0.279   0.274   0.335   0.392   0.65(P)   0.55   0.69(P)   0.232

A (P) after the number indicates that failure was due to surface pitting, not the bending fatigue mode under investigation, and (S) that the test suspended without failure in either mode.

Two questions have to be answered. First, what information, if any, can be abstracted from the results to date for use in new designs? Second, are there any recommendations to be made in respect of the test programme and its future implementation? The reader may wish to make his or her own observations before proceeding.

The first aspect to note is that testing was carried out as the gears were received at only two stress levels. There was no attempt to randomise the allocation to stress level or to test sequence. Any improvement in the manufacturing process will therefore carry forward as a systematic trend. In particular, all first-batch gears were tested at the lower stress level, and the higher stress level contains only second batch gears. Ideally, both the allocation of gears to test and the sequence of testing should have been randomised. Note however, that had this proper procedure been followed, only piecemeal unanalysable results would have now been available!

Acknowledging this restriction, the results can be examined by means of Weibull distributions as originally proposed. Plots of the raw test data are

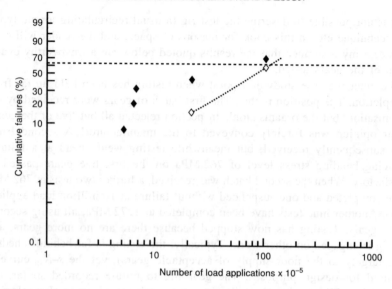

Figure 7.3   Weibull plot of test data at 762 MPa: ◆, original data; ◇, acceptable data; ······ , approximate upper part of a three-parameter Weibull distribution.

given in Figure 7.3 and 7.4. The data relating to pitting, and of course the suspended test, are treated as censored information. Immediately alarm bells should sound. The plot for the lower (762 MPa) stress level shows a marked discontinuity between the third and fourth points (Figure 7.3). This of itself is suggestive of a knee, but the characteristic reversed curvature of the lowest three points, forming the limb below the knee, strongly supports this interpretation. There would appear to be only two relevant failures at this stress level, i.e. failures of sound material. About 30 per cent of the gears tested appear defective, to which must be added those previously rejected on metallurgical grounds and not tested – an appalling state of affairs in my opinion. Indeed, if the sole object of the investigation was to ascertain the cause of gear unreliability then no further work is required. The quality control procedures right down the line need a thorough overhaul. The test data at the higher (1272 MPa) stress level look more promising. All except one point lie on a good, smooth curve. Without the suspicions aroused by the data at 762 MPa this curve would have been accepted.

   Unfortunately, the outstanding point lies just at the knee position (30 per cent cumulative failures) of the previous curve. However, there is no suggestion of reversed curvature at the lower end of this curve, and because the data were all derived from second-batch gears it seems reasonable to treat the data at 1272 MPa as valid. Standard Weibull analysis shows that they can be represented by such a distribution having a

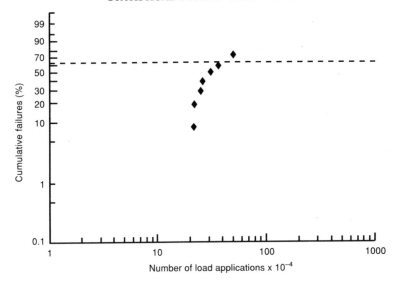

Figure 7.4 Weibull plot of test data at 1272 MPa.

locating constant = 0.2277×10^6 load applications
characteristic life = 0.2211×10^6 load applications
shaping parameter = 0.506

giving a

median value = 0.3349×10^6 load applications.

The correlation coefficient of the least-squares best-fit Weibull straight line is 0.9748 with 5 degrees of freedom. Note: a high coefficient is to be expected, because the data have already been ordered.

What of the test data at 762 MPa? First, a word of warning. It is possible to carry out a standard Weibull analysis using all the data points to obtain a three-parameter function. The correlation coefficient is as high as 0.9824 with 3 degrees of freedom. Superficially, this is acceptable. But the examination of the data point pattern has shown it to be completely invalid. Computer or other automatic analysis of data must be regarded with the highest suspicion, unless a manual appraisal has already shown that it is valid to proceed, or the computer program can recognise a good knee when it sees one. In this case, instead of standard Weibull analysis the following procedure was followed. The limb below the knee ($a_0 > d$) can, as has been shown, be considered as a different failure mode. The lower three data points were therefore treated as censored data, as were the two pitting failures and, of course, the suspended test result. All the data were then ranked (median ranks) in the usual way (e.g. Carter, 1986) to obtain two ranked failure points. Now the upper portion of a Weibull raw data distribution is approximately linear on Weibull axes.

Indeed, one often experiences difficulty in determining the curvature of such plots: see Carter (1986). So a straight line was drawn on Weibull axes between the two acceptable failure points: see Figure 7.3. Note that this is not to be confused with a two-parameter Weibull distribution; it represents the upper portion of an unknown three-parameter distribution. From this line the median life can be very roughly estimated as 8.682×10^6 load applications. It is considered that this is the only useful information that can be deduced from the tests at 762 MPa. We are lucky in that one of the acceptable points is not too far from the median, so any error involved in this very approximate analysis should be small.

It is common practice to assume that the median s–N curve for gears is given by

$$N = ks^m \qquad (7.4)$$

where k is constant and $m = -5$. This is the very well-known Merritt relationship. However, substituting the median values at 762 MPa and 1272 MPa into the general equation (7.4) gives two equations that can be solved for both m and k. The values so obtained are

$$m = -6.3528 \qquad \text{and} \qquad k = 1.7666 \times 10^{25}$$

The value of m is significantly different from the universal Merritt value. It is not clear whether this represents a true difference in the behaviour of these particular test gears or some change in the manufacturing quality between the tests on different batches at 762 MPa and 1272 MPa. In spite of these doubts, the test value is initially used in the subsequent analysis.

There are two methods of estimating a point on any s–N_F curve: (a) using the distribution of life, N, at constant stress, s, and (b) using the distribution of fatigue damage resistance stress, s, at constant life, N. See Figure 6.5.

Method (a) is straightforward. Using the Weibull constants listed above for the higher stress level, $F = 0.1$, and solving for N gives $N = 0.2303 \times 10^6$ at 1272 MPa as one point on the s–N_{10} curve. From the test data available this is the only point that can be calculated in this way. When the proposed test plan is complete, this calculation could be repeated at each stress level to obtain the five points required to define the design curve.

Method (b) is more difficult. It relies on the theoretical relationships, but allows the complete curve to be defined, and provides an independent check on the value deduced by method (a). The procedure is as follows. The generally quoted UTS for 832M13 is 1080 MPa, but the surface tensile strength is about 2380 MPa. Recalling the argument advanced for strain-hardening it can well be claimed that the standard deviations of bulk and surface tensile strengths are the same. Hence the standard deviation can be calculated from the quoted UTS, though we are left with the problem that the value 1080 MPa is not defined in terms of the distribution.

Recall Chapter 3: it could be the mean $(\bar{S})$ or the minimum perceived

$(\bar{S} - 3\sigma_s)$. Most likely it lies somewhere in between; so the calculation will be carried forward using both values as limiting cases. Putting $1080 = \bar{S}$ and substituting in equation (5.7b) gives a coefficient of variation of 0.5, and hence a standard deviation σ_s of 54 MPa. This is the value to use if 1080 MPa is actually the mean. If 1080 MPa is actually the minimum perceived value, the standard deviation is calculated as follows (again using equation 5.7b):

$$1080 = \bar{U} - 3\sigma_s = \bar{U} - 3(0.0564\ \bar{U} - 7.198) \tag{7.5}$$

or

$$\bar{U} = 1326 \text{ MPa} \tag{7.6}$$

and

$$\sigma_U = 67.6 \text{ MPa} \tag{7.7}$$

Assuming that the fatigue resistance is Normally distributed, 10 per cent failures occur at 1.3σ from the mean. So the $s\text{–}N_{10}$ curve can be derived from the median by subtracting 1.3σ from all values of the $s\text{–}N_{50}$ curve. In particular, at $N = 0.2303 \times 10^6$, the median resistance is given by equation (7.4) as 1349 MPa. Hence for the limiting cases

$$\text{upper limit} \quad s_{10} = 1349 - 1.3 \times 54.0 = 1279 \tag{7.8}$$

$$\text{lower limit} \quad s_{10} = 1349 - 1.3 \times 67.6 = 1261 \tag{7.9}$$

Thus according to method (b) the true value of s_{10} lies somewhere between these values at 0.2303×10^6 load applications. The method (a) value was 1272 MPa at 0.2303×10^6 load applications, which falls in the middle of this range. Thus methods (a) and (b) agree with each other, and method (b) could be used for further extrapolation. Note: the above limiting values have been determined solely from a standard median $s\text{–}N$ curve and the quoted ultimate tensile strength of the material. It is therefore a completely independent check on the test distribution derived in method (a).

One further aspect needs to be resolved. The index of -6.353 for the median power curve is significantly different from the universally recognised Merritt value of -5. To clarify this, a fifth-power median curve was made to agree with the measured value of the median at the higher stress level, ignoring all test data at the lower level completely. Repeating the above calculations gave upper and lower limiting values of s_{10} as 1301 and 1287 MPa. The actual value of 1272 MPa lies outside this range. It may thus be concluded that the fifth-power curve is inconsistent with the test data. The experimental value provides consistency, and is therefore retained as a better representation of the material properties of the gears tested.

It is now possible to achieve the object of the investigation and set up an $s\text{–}N_{10}$ curve. It will be based on the test data that have been verified as self-consistent, i.e. a Merrit-type equation incorporating the test values of the constant and the index, and the one firm point derived by method (a) above. This gives s_{10} as 1272 MPa at 0.2303×10^6 load applications. At this life the median damage resistance (s_{50}) is obtained from the revised Merrit equation as 1349 MPa. Hence

$$z = 1272 - 1349 = -77 \text{ MPa} \tag{7.10}$$

According to theory this applies to all values of N, and so, using equations (7.4) and (6.12), the s–N_{10} curve can be written as

$$N = 1.767 \times 10^{25} \, (s + 77)^{-6.353} \tag{7.11}$$

It could be claimed that the objective of the test has been achieved. Indeed, it could be claimed that equation (7.11) has more support than most empirical data! Furthermore, if the design specification required a revised value of F, the appropriate curve could easily be calculated from the same data using the same technique.

However, a serious difficulty remains: it is not possible to say from these tests what proportion of the final total population lie on the lower limb of the knee. Considerably more tests at constant stress would be required than the six originally proposed on the assumption that the population was homogeneous. My own view is that the money would be better spent on improving the quality control of the gear manufacture. Without that, further testing is a waste of time. Meanwhile, confidence in the analysed results could be increased by measuring the UTS of specimens machined from the gears already fatigue tested, to provide the true UTS distribution. This could then be used in place of the one estimated above from general data.

It will be observed that the 'limit' or damage threshold has not been defined by test. In point of fact it is of little consequence in the actual case, because the gears are heavily loaded, and the load distribution is covered by the sloping part of the curve. An additional complication is that, at the longer fatigue lives, fatigue bending is masked or censored by pitting. However, this was not the failure-inducing load for the gears in question at the design life. But to obtain generalised design data, it would be necessary to establish the curve for pitting.

In spite of the many difficulties encountered in this particularly bad case, it has been shown that it is possible to derive the required s–N curve for design – that is, equation (7.11) – if all the information available is used, and the theoretical relationships are exploited to the full. It is also hoped that this case study has highlighted some of the practical difficulties that are not at all obvious in the theoretical treatment, so that they can be minimised by forward planning in similar investigations. A lot of questions still remain unanswered. For example: Is a small order of test gears representative of large batch production? Is the extrapolation of parametric distributions justified? Could one afford to repeat the tests for other materials? Could one afford to repeat the test for all parts of a machine for all failure modes? What about knees? – and so on. There is no unique answer to most of these questions, but they must be considered by designers in relation to their projects.

As an epilogue to this case study, the organisation was obliged to make significant financial cutbacks at that time, most of which were borne by research. They did not consider that they could support a completely revised gear programme, and so no further testing was in fact carried out.

What of load distributions? In most cases there is, by comparison with stress-rupture phenomena, substantial interference between the load and damage resistance distributions. So it becomes necessary to define the whole of the load distribution accurately: it is the many repeated loads of the bulk of the population that do the most damage, and the small contribution of the tail can be less accurately known without introducing serious error into the cumulative damage. This is a very different situation from that controlling stress-rupture failure mechanisms. Because variation in the bulk population is the quantity to be represented, there can be little doubt that conventional statistics is the prime tool. The way ahead is thus straightforward. The real difficulty here is obtaining a distribution that is representative of all users in all circumstances, covering the full life expectancy of the design. That can prove very difficult indeed. It has already been recorded that the difficulties of accommodating a multitude of users, when such are involved, makes the difficulty of accommodating half a dozen suppliers of material one of utmost triviality. But that one has not yet been solved. For this reason I believe accurate specification of the design load to be the greatest difficulty facing designers.

7.1.2 Sensitivity

In assessing possible sensitivity associated with wear failure modes it must be obvious that if only the extreme tails of the load and damage resistance distributions are interfering then tail sensitivity problems similar to those encountered with stress-rupture modes must arise. Equally obviously, this situation can come about only when the damage resistance has a finite threshold, for if the threshold is substantially zero, strong interference must take place. Figure 4.4 may be used as a guide to identify possible sensitive regions, reading damage margin for safety margin.

As a more direct illustration, in Figure 7.5 the life to 10 per cent failures has been plotted against the damage margin for a loading roughness of 0.7, when both load and strength are Weibull distributed. For this purpose the damage margin is defined in an analogous manner to the safety margin as

$$\text{Damage margin} = \frac{\bar{E} - \bar{L}}{\sqrt{(\sigma_E^2 + \sigma_L^2)}} \qquad (7.12)$$

where $\bar{E}$ is the mean value of the damage threshold, and σ_E is its standard deviation. Negative values of the damage margin are not uncommon. The

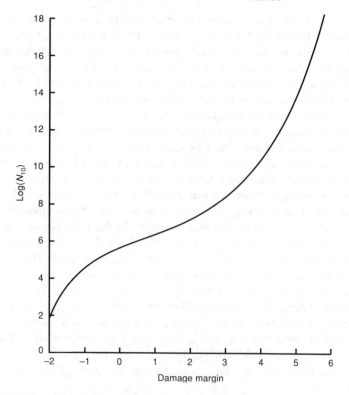

Figure 7.5   Variation of life with damage margin. Loading roughness = 0.7.

loading roughness to be used in this case is strictly

$$\text{Loading roughness} = \frac{\sigma_L}{\sqrt{(\sigma_E^2 + \sigma_L^2)}} \qquad (7.13)$$

though because σ_E is roughly equal to σ_S, the loading roughness has more or less the same value for most failure mechanisms.

Figure 7.5 shows that the sensitive region corresponds to very high numbers of load applications associated with high damage margins, i.e. when the tails are interfering. On the principle that sensitive regions should be avoided for high-quality designs, the same argument that was used for stress-rupture mechanisms suggests that all damage should be avoided in these circumstances. This implies that a stress-rupture approach should be used. It was the underlying reason for doing so when estimating compressor blade factors of safety in section 5.2. Such design need not concern us further here.

Although gross sensitivity may be recognised and mitigated to some extent by such techniques, the reliability achieved at any life, or the life for any specified reliability, is still very dependent upon the design parameters, even

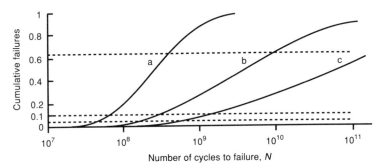

Figure 7.6   Life distributions at various loadings. All loads are Weibull distributed, and have same mean and standard deviation. (a) Shap-
ing parameter = 2.0; (b) shaping parameter = 3.44 (near normal);
(c) shaping parameter = 5.0.

in the insensitive region. There are so many combinations of load and strength
distributions and shape of s–N curve that no generalisation is possible. Each
case must be considered in its own context, though gross load–damage interference and steeply sloping s–N curves make for insensitivity. As an illustration of the sensitivity that may be encountered in the 'average' design, Figure
7.6 shows the probability of failure plotted against age for different Weibull-
distributed loads. The mean and standard deviation of each load is the same,
so superficially the same load is applied. However, the shaping parameter has
been chosen to give a roughly Normal distribution for one load, and right and
left skewing to the others (the first achieved with Weibull shaping parameter
of 3.44, the latter with 2 and 5 respectively). The difference from the 'Normal'
could well be that applied by different operators. Figure 7.6 shows that the N_5
and N_{10} lives for the skewed distributions differ from the Normal by a factor
of about 5. For the N_{50} the life ratio is about 10. Although not high in comparison with values pertaining to truly sensitive situations, these values would
still be regarded in an absolute sense as very high for practical purposes. For
example, a product that required four or five times the expected number of
replacements (such as scheduled replacement at N_5 or N_{10}) would not be received with acclaim by the customer!

There is, unfortunately, another more insidious form of sensitivity that can
arise in some wear processes. It is particularly associated with s–N curves
that have a low slope, though aggravated by a high standard deviation of the
damage resistance. Even with precisely defined load distributions, items of
various resistances from the same population can have markedly different lives.
The result is an extended, i.e. sensitive, life distribution. The curves of Figure 7.6 show that for a given load the ratio of long life to short life can be
100:1 or even 1000:1. This is confirmed, for example, by the test data in
Figure 6.11 or 6.14, which show life ratios of 100:1 at the same stress level.

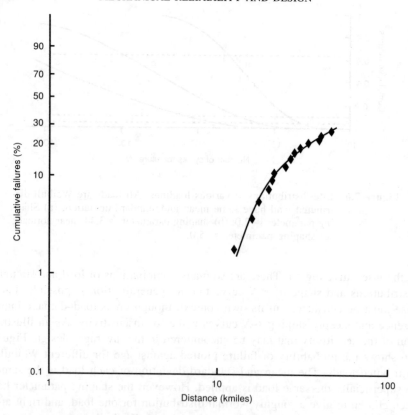

Figure 7.7 Failure pattern of water pump (Weibull axes). Locating constant = 1180 miles; characteristic life = 170 000 miles; shaping parameter = 0.8.

As an actual example, Figure 7.7 shows the life to failure of a conventional water pump fitted to an ordinary petrol engine, though this is by no means an extraordinary example. The data have been plotted on Weibull paper, and a distribution has been fitted. The lives plotted are to first failure, so no maintenance aspects are involved. The significant features of the distribution are: (a) there are no failures up to 12 000 miles; (b) the median life is about 100 000 miles, which can also be taken as the life of the parent equipment; and (c) the mean life is about 220 000 miles. The actual load and damage resistance distributions for this example are, of course, unknown. If an item of 220 000 miles average life is being discarded at 100 000 miles (on average), about half the population's useful life is being thrown away (ignoring any cannibalism): hardly a mark of effective design. It is of course possible to move the distribution bodily along the age axis at the design stage by adjusting the damage margin. However, any adjustment towards lower lives in order to make better use of the pump's upper life distribution must entail a reduction in the failure-free life. In this example, higher warranty costs would probably be incurred. On the other hand, any attempt to increase the warranty period by increasing the failure-free life must incur a loss of the pump's overall life potential. A balance must be struck. This aspect is pursued in Chapter 10 on design strategy. The problem can be reduced by selecting materials of lower variability and steeper s–N curves, but even when all this is done, wide life dispersion (sensitivity) remains a serious problem. Extended distributions are curtailed by high loading roughness, but this is a parameter usually outside the designer's control.

7.1.3 Miner's rule

Errors arising from the Miner–Palmgren rule are much more difficult to assess. It is usually claimed that there are two sources of inaccuracy. In the first place, the damage done may not be a linear function of the number of load applications to cause failure. Second, the proportionality may differ between virgin and damaged parts, or between parts of different damage. The sequence in which variable loads are applied becomes a parameter in any life assessment. The sequence is, however, itself a stochastic quantity.

Virtually no work has been done on the validity of this rule other than for fatigue. For this particular failure mechanism, Miner's rule is often written as

$$\sum \frac{n_i}{N_i} = G \qquad \text{at failure} \tag{7.14}$$

in some acknowledgement that the basic rule expressed by equation (6.5) is an inadequate model. The Miner summation, G, then becomes some empirical number other than unity. Shigley quotes

$$0.7 < G < 2.2 \tag{7.15}$$

though I have seen values outside this range quoted in the literature. It is not clear to me what G is supposed to represent. It may cover the two unknowns noted above, but some designers have told me that they use a low value of G – usually about 0.8 – 'just to be sure'; i.e. G is being used as a disguised factor of safety. This cannot be acceptable. I have seen no attempt to correlate G with design parameters, apart from some claims that it allows for stresses above the yield.

In due course, crack-propagation-based theories will provide the definitive techniques for dealing with distributed loads. Unfortunately, they still have some way to go. Realistically, designers are at present stuck with Miner, and have to make the best of a bad job. What then is the best way to implement it?

According to modern theory, whatever loads are applied to any item of a population, microstructurally short cracks are always formed and will grow until they reach length d, at which length they will be arrested unless transformed

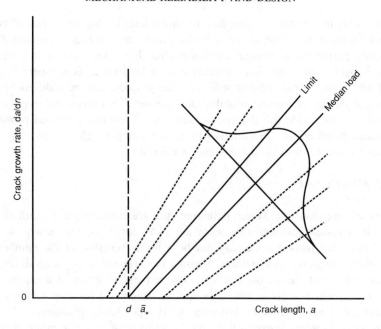

Figure 7.8   Physically small crack growth for an item of a population subject to a distributed load.

into a physically small crack. Note: all loads, even if less than the limit, will
be contributing in some measure to the microstructurally short crack growth,
and hence to the 'damage'. This contribution is ignored in the conventional
Miner summation when the load is less than the fatigue limit.

Suppose that a deterministic load, such as the mean representing a distributed load, is less than the limit condition of an arbitrary item from a product
population to which it is applied. In crack propagation terms this implies that
$\bar{a}_* > d$ for the item in question, where $\bar{a}_*$ is the value of a_* for the mean
load. The situation is illustrated in Figure 7.8 by the full lines. The minimum
physically small crack length is then greater than the maximum microstructurally
short crack length d, and so a physically small crack cannot form. The life of
the item would be infinite. Now let the load be distributed, i.e. the real situation. In addition to the full line in Figure 7.8 (or the deterministic version in
Figure 6.8), there must be other physically small crack lines on the diagram,
each corresponding to the magnitude of an individual load from its distribution. Some of these are shown dotted in Figure 7.8. Under the influence of
those individual loads whose $a_* < d$ (the two dotted lines on the left of the
distribution in Figure 7.8, for example), the microstructurally short crack can
transform into a physically small crack. So a physically small crack will develop and grow. This does not conflict with contemporary ideas, because these
individual loads are greater than the limit, and would be included in a conventional Miner summation. As the physically small crack grows under the

influence of the high loads, it will eventually reach length $\bar{a}_*$. It will be seen from Figure 7.8 that the mean load represented by the full line in that figure, which is less than the limiting value for that item, will once again be contributing to the crack growth. However, by contrast the load will be considered to be doing zero damage according to conventional ideas, and so will not be included in the Miner summation. As the physically small crack length a increases, the physically small crack propagating condition, $a > a_*$, will be satisfied by more and more of the lower loads, which will then also contribute to the crack growth. In fact, all loads in the distribution will eventually do 'damage', and must be included as appropriate in an effective Miner summation. The life cannot be infinite with a distributed load unless it is intrinsically reliable with respect to the limit.

It would appear that neglect of the damage done in both crack regimes by loads below the fatigue limit is the most likely source of empirical values of the summation G being less than unity.

In the general case of load-damage resistance interference, then every load in the distribution will be contributing to the damage except when the crack length a falls in the range $d < a < a_*$. Unfortunately, the range cannot be calculated (at present), and approximations have to be made in any practical design calculation. Thus in evaluating the conventional Miner summation, damage done by loads below the limit is ignored, even though – as we have seen – these loads actually do cause damage. An alternative approximation is to ignore the interval $d < a < a_*$: i.e. ignore the limit, where the limit is interpreted in terms of modern crack propagation theory. An effective s–N curve to meet this approximation is obtained by extrapolating the sloping part of the s–N curve to higher N. An extended Miner summation is then evaluated using this curve. Failure always occurs when the summation equals unity under this approximation. Obviously, the conventional Miner summation can overestimate the life to failure, while the extended summation can underestimate the life. If the whole of the load distribution is encompassed by the sloping part of the s–N curve, then both methods give the same result. If the load distribution extends below the limit, however, a difference between the two summations is introduced, which requires a value of G less than unity to be applied to the conventional summation to obtain agreement.

The above account of accumulating damage would certainly not support values of G greater than unity. The explanation advanced by Talreja (1979) to explain some of his results may then apply. It is well known that crack growth under substantially deterministic loading can be delayed by the application of single or multiple overloads. The exact mechanism appears not to be fully understood at present, but it is believed that large residual stresses produced ahead of the crack tip by the overloads play an essential role in the mechanism. Talreja has suggested that the largest stresses in a distribution sometimes act as such overloads, and prolong life. It is a reasonable suggestion. Unfortunately, it has disturbing consequences. In an extreme case an operator who 'improves' the loading by curtailing the right tail of the load distribution could actually shorten the life of the product! When is a high load a constituent of a distribution and when is it an overload? So much is not known about the various aspects of empirical G values that their use is more an act of faith than of reason. It must be considered very unwise to use them in design unless they are supported by extensive field (not test) evidence, and by data from a representative sample of independent operators.

To sum up: it would appear that the designer has three options in the application of the Miner-Palmgren rule. He can use any of:

1. the basic Miner summation ($G = 1$);
2. an empirical Miner summation ($G =$ empirical value);
3. an extended Miner summation (unlimited s–N curve; $G = 1$).

The preceding discussion could act as a guideline in selecting the most appropriate option. At the same time, the designer must consider the possibility of loads other than fatigue (e.g. corrosion) enabling the crack to bridge the d to a^* gap. That situation is best represented by option 3.

If G is not taken equal to unity then equation (6.27), which has been derived on the basis that it is, must be modified to read

$$N = \frac{G}{\int_0^\infty \frac{L(s)}{\zeta(s + z)} ds} \qquad (7.16)$$

Perhaps overmuch attention has been paid to Miner in this book, largely because it is so widely discussed. Indeed, I find protagonists can become quite emotional on this topic. My own view is coloured by the knowledge that the errors associated with Miner are small compared with those from other sources. For example, Miller (1991) has shown that in what he admits are extreme examples the Miner estimate can be in error by a factor of 2 when contrasted with crack propagation estimates. By comparison it has been shown that feasible changes in the load variation can give rise to errors of 10:1 in the median life (Figure 7.6), and this would be very much higher indeed if possible changes in the mean load were to be included. Even when the load and strength are accurately known it has been shown that differences in life near 1000:1 can arise between members of the same population. It is against such figures that the designer must judge Miner. I would suggest that at present it is far more important to obtain comprehensive data on distributions than to 'dot the i's and cross the t's' of Miner estimates.

7.1.4 Review of difficulties

Reviewing all these difficulties – distributions, sensitivity, and Miner – the most serious obstacle to a design methodology based on statistically defined input data is ensuring that the load distribution covers all eventualities. But no method of design can succeed without a clear knowledge of the load, so this difficulty should not be held as a debit solely against statistical methods. At least for the moment that problem can be set aside as a common problem. In that case a design methodology based on statistically defined inputs becomes feasible for wear modes. Before reaching any firm conclusion, however, quantitative comparisons must be made between results from existing empirical design practice and those based on statistical data.

7.2 Comparison of statistical and empirical design

There is in fact a good deal of similarity between the statistical design method described in the last section and some contemporary empirical design practices. Both may be said to be based on a 'safe' s–N curve, which has been derived in some way from the standard s–N curve. The essential difference is an empirical curve based on factors contrasted with a statistically defined quantitative s–N_F curve. Thereafter the design procedures are the same. So how do the differently estimated 'safe' s–N curves compare? This is the critical test.

Most empirical methods are of course the private preserve of those practising them. However, turning to one of the most sophisticated branches of engineering, we are fortunate in having a very good description of the philosophy behind the fatigue design procedure for aircraft structures in the UK (Cardrick et al., 1986). A 'safe' s–N curve is derived from the median curve by empirical factoring, but it has been found that a single factor is inadequate. Consequently, two factors are used. For high-cycle fatigue (near to and including the limit) the stress is factored to obtain part of the safe s–N curve from the standard median curve, but for low-cycle fatigue the life is factored to obtain another part of the safe s–N curve. Because these factored curves do not intersect, a blend from one to the other, which avoids any inflexion, is used for intermediate cycles. Figure 7.9, which has been adapted from the paper by Cardrick et al., shows how it is done, using typically a factor of 1.37 on stress and 3 on life. The safe s–N curve is then used in design without any further factor of safety, though an additional safeguard is introduced by the condition that a 20 per cent increased load should be sustainable for at least half the design life. The safe s–N curve may be considered 'a bit of a dog's breakfast', having two differently based empirical factors and an arbitrary blend, the latter coinciding with the central load in many application (as in Figure 7.9 in fact); but one has every sympathy with those responsible for its derivation in an environment where safety is paramount. The factors are based on extensive historical experience, as detailed in the paper, and are the result of a systematic development of the factor method. This safe s–N curve is probably the most trustworthy, and certainly the most documented, of the empirical techniques. It is of course not formally stated how safe 'safe' is,

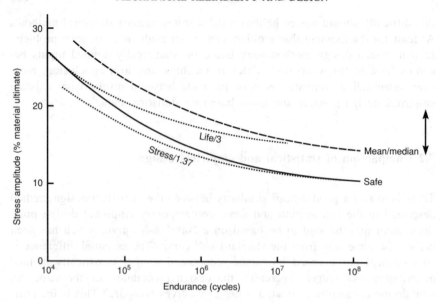

Figure 7.9   Actual and 'safe' *s–N* curves of a typical aluminium alloy; 'mod-
                 erate' notches. Mean stress 30% of material ultimate. Arrows
                 indicate range of amplitudes of most importance for combat aero-
                 plane wing bending. After Cardrick *et al.* (1986).

and this constitutes the major difference between the empirical and statistical
approaches. It is, however, stated elsewhere in the paper that 'the chance of
a failure at full life under anticipated loads should be about 1 in 1000'. That
figure will be assumed to apply in the comparison now to be made.

   The use of the empirical *s–N* curve is subject to 'a requirement to design
structural details which have a good fatigue life that is not unduly sensitive
to increases in the severity of the anticipated stresses and abnormal defects
arising in manufacture or service'. Abnormalities are stated to 'include discrete defects in material or manufacture, or scratches, cuts, impact damage
and the breakdown of protective coatings'. With these conditions, plus the
elimination of assignable causes of variation by the strict quality control associated with aircraft manufacture, it would seem reasonable to assume the
absence of knees in the fatigue characteristic of aircraft structural materials,
and a roughly Normal distribution of fatigue resistance. In that case the 1 in
1000 chance quoted above would correspond to a stress about 3 standard
deviations below the mean damage resistance. It is also stated in the reference that the coefficient of variation of the fatigue limit is about 10 per cent
for the structural details with which the paper is concerned. It is not clear
whether this applies to the material data from which Figure 7.9 was constructed; neither is there a true limit to the curve on that figure. Assuming
that the limit can be taken at 10^8 cycles, the fatigue limit of this material is

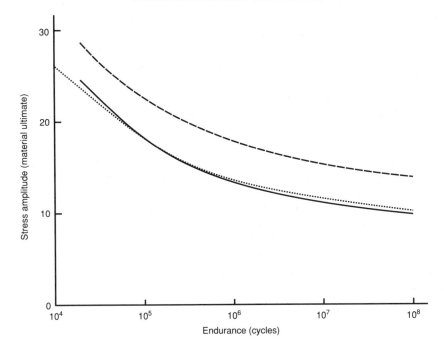

Figure 7.10 Comparison of empirical 'safe' s–N curve and calculated s–$N_{0.1}$ curve: – – –, median curve used for both empirical 'safe' and calculated curves; ——, calculated curve, ·······, empirical curve.

14 per cent of the ultimate. Using the coefficient of variation of 10 per cent, the standard deviation of the fatigue limit (damage threshold) is 1.4 per cent of the reference ultimate tensile strength. It must be the deterministic mean or median value of the UTS that is being used as a reference, so because the standard deviations of the limit and the ultimate are theoretically the same, this is also the coefficient of variation of the ultimate. It seems low to me. Nevertheless, this, in conjunction with the 3σ location of the safe s–N curve, gives a value of z as

$$z = -3 \times 1.4\% = -4.2\% \text{ UTS} \tag{7.18}$$

where z is defined by equation (6.11) and also in figure 6.18. Hence subtracting 4.2 per cent from the non-dimensionalised stress of the median curve in Figure 7.9 gives the calculated 'safe' s–N curve (actually the s–$N_{0.1}$ curve). It is shown as the full line in Figure 7.10, where it is compared directly with the empirical safe curve (dotted) taken from Figure 7.9. The curves are the same to all intents and purposes.

Indeed, the comparison is extraordinarily good, particularly considering that the F value is well outside that which could be readily measured on any one material, and it behoves us to stand back and assess its true worth. In the first place, the acceptable levels of failure and the coefficient of variation used in the calculation are clearly rounded global values. It is on this general level that the comparison has to be made, not on the specific material that the theory would claim also to model. Nevertheless, this need not be all restrictive. It is first claimed the the empirical 'safe' curve must also be an s–N_F curve (F constant). It would make a complete nonsense of the empirical approach if this were not so; F could be interpreted as any F value to fit the factors! Second, the coefficient of variation applies to the limit, i.e. to only one high value of N. So even if the agreement at this N were coincidental when comparison with a given material is being made, as in Figure 7.10, it could not be claimed that the agreement at the lower values of N were so too. It is stretching the bounds of credulity too far to suggest that the agreement of the two curves in Figure 7.10 is coincidental *over the whole N range*, particularly when the factors defining the empirical curve are based on two different quantities – stress at high N and life at low N. It would thus seem safe and fully justified to conclude from this comparison that the statistically derived curve in Figure 7.10 is every bit as good as the empirical curve, and is indeed to be preferred for the following reasons:

1. It has a physical significance: it is a defined constant F (or R) curve and can be calculated for any numerically specified value of F; it is not an unspecified 'safe' value.
2. It can be calculated accurately over its whole N range using only measurable properties of the material.
3. It has theoretical backing.
4. It is simple.
5. It allows sensitivity to be checked by varying F about the target value.
6. It is fully transparent, and can be subject to a genuine audit.
7. The empirical factor method is fundamentally flawed.

The last assertion may require justification. The empirical factor based on stress can be written as

$$\text{Factor} = \frac{\bar{E}_N}{\bar{E}_N + z} = \frac{1}{1 + z/\bar{E}_N} \qquad (7.19)$$

where z is numerically negative when it defines the safe curve. It has been demonstrated that z is constant for a given value of F for all N. But $\bar{E}_N$ varies with N according to the s–N curve. Hence it is impossible for the factor to remain constant for all values of N. To overcome this deficiency of the factor, the empirical design of aircraft structures is based on two factors, one at either end of the range. It is only a matter of convenience and historic precedent that one is based on stress and the other on life, but it leaves the middle portion of the curve undefined. More (five?!) factors would be required if the 'safe' curve was to be defined to the same accuracy as current s–N curves. An empirical factor can have no unique value or significance.

As with stress-rupture, one needs to treat agreement between different methodologies with caution; neither may represent reality. The Normal distribution has been assumed in calculating the statistical curve, and the reader will recall doubts cast on this distribution in connection with stress-rupture of aircraft structures. Unfortunately, Cardrick *et al.* make no reference to or comment on the distribution of their fatigue properties. It would be a remarkable coincidence if a rectangular distribution with a tail were behaving as a Normal distribution. Gordon's rectangular distribution applies to the strength of completed structures, of course, whereas the above comparison is concerned with the damage resistance of the basic material, and the reservations may be irrelevant. It just leaves a lingering doubt.

It should also be recalled that evaluating the appropriate s–N curve is only half the battle. The load distribution has to be determined, and must represent field practice. Then the controversial Miner's rule has to be used to calculate the life for the target reliability. Cardrick *et al.* state that 'no further factor will be needed when Miner's rule is adjusted for local residual stresses', and the Miner summation is based on the local conditions at the root of the stress concentration. I would not dissent from any of that. It may be worth recording, however, that in their example the load distribution is encompassed by the sloping part of the s–N curve: see Figure 7.9. According to the discussion of Miner's rule in Chapter 6 this conclusion could be expected, but may not be universal.

The reader may have deduced from the case study contained in the previous chapter that some organisations have moved beyond the aircraft industry's position, in that they are measuring s–N_F curves, but because they do not publish any information or results, no assessment is possible. However, I do not know of any organisation using them as a primary design tool.

In conclusion, then, it would appear that methods substantially the same as that described in the first section of this chapter are being used, most probably alongside conventional empirical methods, with what results cannot be said. In the one direct comparison that *is* possible, the empirical and the statistically based methods are in very close agreement. For reasons already given, the method based on statistically defined inputs is to be preferred.

References

Cardrick, A.W., Maxwell, R.D.J. and Morrow, S.M. (1986) How future UK military aircraft will be designed for tolerance to fatigue damage, *Conference on Fatigue of Engineering Materials and Structures*, Vol. 1, Institution of Mechanical Engineers, London, p. 15.

Carter, A.D.S. (1986) *Mechanical Reliability*, 2nd edn, Macmillan Press, Basingstoke.

Miller, K.J. (1991) Metal fatigue – past, current and future, Twenty Seventh John Player Lecture. *Proceedings of the Institution of Mechanical Engineers*, **205**, 1–14.

Talreja, R. (1979) On fatigue reliability under random loads. *Engineering Fracture Mechanics*, **11**, 717–732.

8

A design methodology

It may seem ironical that a design methodology based on statistically defined inputs is to be preferred when complex wear processes are involved, but is of doubtful advantage for the much simpler stress-rupture mechanisms. The reason for this must be fully recognised. A lower level of design reliability is acceptable for wear because it can be countered by maintenance (when properly carried out), while maintenance can have no effect at all on the stress-rupture processes. Thus for aircraft structures, for which a very high reliability might be thought a *sine qua non* on account of safety, a full life design probability of fatigue failure as high as 1 in 1000 is considered adequate – though this is not of course the overall probability of failure. Consequently, for design involving wear processes only the accessible part of the probability density curve is generally required to provide the statistical input data. The superiority of design based on statistically defined inputs then becomes overwhelming – as one would expect. For stress-rupture an inaccessible part of the probability density curve is required for an accurate quantitative solution. In such circumstances statistical methods are not invalid, but they are impotent. So too are the empirical 'factor' methods. Techniques that claim to access the inaccessible are just crystal balls. Added to which, it has been shown that 'factors' are fundamentally flawed, in the case of both stress-rupture and wear, even when applied to the accessible part of the curve. In these circumstances the obvious solution must be to use the superior, fundamentally sound, i.e. the statistically orientated, method whenever it is workable, and, as offering a consistency of approach, also when it is still fundamentally sound but less tractable. The balance of advantages and disadvantages for either approach to stress-rupture is so evenly balanced (Section 5.3) that it cannot influence the decision in any realistic way. The option is even more justified because the vast majority of design activities are associated with wear, for which tractable solutions are possible.

There is in fact a good technical reason why statistically defined inputs should be used for stress-rupture design, once it has been chosen as the preferred method for wear: it greatly facilitates design involving both static and dynamic stresses. Figure 8.1 shows what is usually called the the **modified**

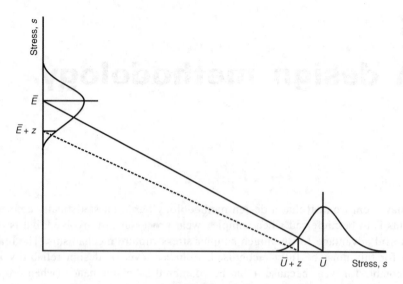

Figure 8.1 Modified Goodman diagram when both ultimate tensile strength and damage threshold are distributed. $z = s_F - s_{50}$.

Goodman diagram used in deterministic design. In addition, the distributions of the fatigue limit and the ultimate tensile strength have been superimposed. Now it has been shown that, as a consequence of the near-deterministic nature of strain hardening, these distributions have the same form and variation about their means. There is a one-on-one ordered correspondence between the individual items making up the two distributions in Figure 8.1. It would be perfectly valid, therefore, when designing for $F\%$ failures to join the ordinate and abscissa for those $F\%$ failures as shown by the dotted line in Figure 8.1, in the same way that the 50 per cent failure points are joined for deterministic design. The dotted line is then the Goodman line for $F\%$ failures. It can be used in exactly the same way, without factors, as the standard Goodman line is used, with factors, in contemporary deterministic design. Clearly, exactly the same remarks apply to the Sodderberg line, if that is used, because the variation of the yield is the same as the variation in the ultimate tensile strength. For the same reason, the yield limiting line that is often added to the modified Goodman diagram can be constructed by joining the ordinate and abscissa for $F\%$ yielding. In principle, one is carrying out exactly the same operation for the $F\%$ failure value as one would in using the 50 per cent value in conventional deterministic design. One then obtains an $F\%$ (or $R\% = 100 - F\%$) design without using factors. This assumes, of course, that intrinsic reliability is maintained up to the life for $F\%$ failures. It is worth emphasising that this simple procedure is only possible on account of the one-on-one ordered relationship between the distributions. Failure to recognise this leads to a much more complicated and unwieldy statistical approach sometimes found in the literature. It is both unnecessary and an incorrect model of the material behaviour.

The case for a design methodology based on statistically defined input data becomes thus almost irresistible, even if inaccessible parts of load or strength distributions require some empiricism, such as when very high reliability is demanded from a wear mechanism, or a stress-rupture mechanism is involved. Although in some ways unfortunate, it is inescapable. Even so, the statistical approach allows the risk associated with the empiricism to be assessed. Empirical rather than theoretical values of k_S and k_L would be used. But theory still gives a lead. If the bulk of a load or strength distribution is well represented by a Normal distribution, for example, then something strange is going on if k_S or k_L were far from 4.75. It would warrant further attention. Likewise, if the bulk of a load distribution is well represented by a Weibull distribution with a shaping parameter of 2 (a common situation), then k_L should not be far off 6.1. There may be good and valid reasons for differing values – but there may not!

Once loads and strengths are statistically defined, the whole of design can be quantified – and audited. There is no need to make use of load–strength interference modelling or any other esoteric epiphenomenological statistical methods in actual design. The relationships between the material properties make them superfluous. Of course, the blind application of any methodology cannot be recommended. A proper understanding of the underlying theory and principles involved is absolutely essential if all the pitfalls are to be avoided. I would also have thought that a qualitative understanding of load–strength modelling, as a means of appreciating sensitivity, is essential.

In fact no design methodology can ignore sensitivity. The requirement that aircraft structures should have 'a safe life that is not less than half the specified life with loads elevated by a factor of 1.2' is really a sensitivity safeguard. But because factoring has been discredited, it is of doubtful expediency. Furthermore, the factor of 1.2 has been adopted because 'many details are found to experience stresses that are 10 per cent or so higher than those anticipated in design; occasionally stresses increase by 20 per cent or more'. If this is so, then in any statistical representation such stresses should have been included in the load distribution – at the frequency where they were found to arise. To do otherwise is statistically unsound. It would, however, be perfectly valid to regard those items of a product in the hands of different operators as different populations subject to different load distributions.

Statistically, sensitivity can be measured by **confidence limits**. These limits define values of the life, N_F, within which we have a given confidence that the true value lies. The given confidence is usually taken as a symmetrical 90 per cent. The limiting values of N_F can be deduced from the limiting load distributions in the same way as the expected life, N_F, is calculated from the expected load distribution. In deducing the expected load distribution from field data, median (50 per cent) ranks are usually used; in deducing the limiting distributions, 5 per cent and 95 per cent ranks are used (for a symmetrical 90 per cent confidence limit). The difference between these two lives, or either of them and the expected life, is a measure of the sensitivity of the design.

One feature of this calculation must be emphasised. The 5 per cent rank load distribution is not the worst-case condition; it is purely a statistical concept measuring the extreme distribution to be expected, at a defined confidence level, from the known data. There is no way in which we can evaluate an excessive load that may be placed on any product, if that load has not been experienced beforehand with a similar product in the field or with prototypes on trial, because such loads need not be related in any way to the central load.

As an alternative to assessing sensitivity as the change in life at a given reliability consequent upon a load change, sensitivity can be assessed as the change in life for the same load distribution when the reliability is changed to some limiting value. It may not measure what we are really interested in, but it is a more realistic (and simpler) calculation.

It may be worth recalling at this point that the design target reliability is the specified operational reliability discounted for any on-condition maintenance and fail-safe features: i.e.

$$(1 - R_{op}) = (1 - R)(1 - M)(1 - R_{fs}) \tag{8.1}$$

where M is the probability of success of the on-condition maintenance (maintainability) and R_{fs} is the reliability of any fail-safe devices; R is the target design reliability to achieve the specified operational reliability R_{op}.

We are now in a position to summarise the steps to be taken in a design based on statistically defined inputs. It is assumed that a preliminary design has already been carried out, using the usual rough-and-ready estimates and rules of thumb that are always adopted in the early stages. The following actions are then required:

Actions common to both wear-out and stress-rupture:

Step 1 Select the most appropriate material for the part in the normal way, and define its mechanical properties in statistical terms. Are these based on trustworthy sources? If not, seek confirmatory evidence or put necessary tests in hand.

Step 2 Define statistically the design load cycle or distribution, both that applied by the operator and that applied internally by other parts and components.

Step 3 Are there any features, such as multiple suppliers or others discussed in Chapter 3, that would demand modification of the load or material property distributions? If so, what are their probabilities?

Step 4 Do the material data apply to the 'as finished' condition? If not, what additional variation of the known properties will be introduced during manufacture?

Step 5   Modify the basic distributions to conform with the 3 and/or 4.
Step 6   Locate the failure-inducing stress, using finite element or other tech-
         niques, and estimate its probability density function, $L(s)$, from the
         load distribution at step 5.
Step 7   What is the reliability of any fail-safe device?

Actions specific to stress-rupture:

Step 8   Select appropriate values of k_S and k_L for distributions derived in
         step 5:
         (a) using values based on experience (if available);
         (b) using theoretical values corresponding to distributions at 5 and
             6.
Step 9   If values at (a) and (b) differ significantly, account for the differ-
         ence.
Step 10   Estimate required strength to meet worst-case condition using equation
         (4.11).
Step 11   What is the probability of any fail-safe device or technique not
         working? Revise design of fail-safe device if required.

Actions specific to wear-out:

Step 8   What is the probability of the part being accidentally damaged by
         impact etc.:
         (a) during assembly into the parent equipment?
         (b) during any maintenance or other repair activity?
         (c) during transport or other handling operation?
         (d) by improper use?
Step 9   What is the probability of any part made from the specified mater-
         ial having a knee or other similar characteristic?
Step 10   What is the probability of any maintenance procedure failing to
         detect and rectify at specified intervals defined wear damage?
Step 11   What is the allowable probability of failure, F, attributable to the
         relevant wear mode in a well-finished homogeneous product that
         meets the customer specification at the specified life allowing for
         defects under 7, 8, 9 and 10? This is obtained using equation (8.1),
         the product rule, and equation (7.3) as necessary.
Step 12   Has it been established that the material specified at step 1 con-
         forms to the behaviour characterised by constant strength followed
         by sudden death, at least to the extent that intrinsic reliability is
         achieved up to the design life for the value of F derived in step 11
         to avoid premature stress-rupture failure modes? If not, this may
         be the critical design criterion. Consider a more trustworthy mater-
         ial or a stress-rupture (no damage) form of design.

Step 13   Deduce the s–N_F curve, either:
  (a) preferably directly from test data without the use of parametric distributions; or
  (b) using parametric distributions, if only limited test data are available; or
  (c) from the conventional s–N curve and the standard deviation of the UTS or the yield as in figure 6.16; if (a) and (b) are not possible, and as a check on (a) or (b) if one of them is.

Step 14   Check the damage resistance distributions at constant s and constant N resulting from step 13 for consistency. If inconsistent, further test data are essential.

Step 15   Decide which Miner summation is most appropriate.

Step 16   Calculate the life, N_F, for $F\%$ failures using the relevant equation (6.27) or (7.17).

Step 17   If the calculated N_F is unsatisfactory or unacceptable, consider:
  (a) redesign with new dimensions or new geometry;
  (b) revised maintenance (inspection) intervals;
  (c) better quality control to eliminate or minimise knees, if present;
  (d) better quality control to reduce the probability of accident damage noted under step 8;
  (e) higher maintenance efficiency – note that this is a doubtful route unless different proved techniques are available;
  (f) alternative materials.

Step 18   Check sensitivity:
  (a) for a more damaging load or reliability;
  (b) for extended life at design load.
  If unacceptable, consider:
  (i)   better quality control;
  (ii) an alternative material;
  (iii) eliminating wear by a form of stress-rupture design.

The above steps are not intended to be a definitive procedure for slavish adoption in design offices, but rather a general indication of the ambience that must surround design for quantified reliability. Actual procedure has to be controlled by the practices of the designer or design team and the circumstances surrounding each undertaking.

Although in Chapter 3 we set out to establish only a qualitative understanding of the design method, in fact some very positive conclusions have finally been reached and a workable design methodology formulated, tempered only by the lack of data that must ultimately fustrate any methodology. Provided that data limitations and sensitivity restrictions are always recognised, design for a quantified reliability becomes feasible. Even so, it must be appreciated that these restrictions will limit an assessable design reliability to at best 1 in 1000 chance of failure. Still, this may be regarded as covering a very substantial proportion of run-of-the-mill designs. Design for higher quantified reliabilities becomes problematic. A degree of intelligent anticipation or even guesswork must be called upon. For safety requirements it is possible to design for a nominal 100 per cent reliability – though pedantically an unknown reliability – by means of intrinsic reliability with respect to the strength for stress-rupture modes or with respect to the damage threshold for wear modes, provided that the weight penalty can be accepted. As we have already seen, however, this is by no means straightforward. As an aside in this connection, I have doubts whether specifications calling for a 1 in 1 000 000 or 1 in 10 000 000 or so chance of failure have any real meaning so far as mechanical reliability is concerned. Such extreme quantification must be treated more as wishful aspirations than actual demands, even though these high (and higher) reliabilities are feasible – achieved by gross overdesign to accommodate data uncertainty and sensitivity problems. As an alternative, tractable reliabilities can always be enhanced by fail-safe techniques or on-condition maintenance. So *in toto* the situation is not altogether bleak.

The close similarity between the above methodology and current practice has already been noted – incidentally satisfying a major objective of the book, as set out on page 8. Hence in the next chapter we shall look at some conventional supporting techniques and discuss their relevance to a design methodology based on statistically defined input data.

9

Miscellaneous supporting techniques

There are several supporting techniques that are often advocated for conventional design. This chapter examines their application to design based on statistically defined inputs. No attempt will be made to describe these techniques in detail, it being assumed that the reader is familiar with their conventional use. The techniques covered are checklists, FMEA and FTA treated together, and design reviews and design audits, also treated together.

9.1 Checklists

Checklists will be familiar to all designers. They are simply a list of all relevant failure modes or mechanisms, and act as reminders to ensure that the design has been assessed as adequate to meet all possible circumstances. It is a very powerful tool, being the only means I know by which possible failure modes can be identified. Such mechanisms as vibratory stress that has to be accommodated or eliminated, fasteners that can become unfastened, potentially corrosive environments, seals that do not seal, unacceptable tolerance stacking, inadequate provision for maintenance activities, and so on, all spring to mind in the quiet of one's study, but can be overlooked in the hurly-burly atmosphere of the design office. A typical (but incomplete!) list has been presented by the author in an earlier work (Carter, 1986). In that list associated failure modes have been grouped together, but the grouping is one of convenience. It has no fundamental significance. Such a checklist is little more than a glorified 'shopping list'. And it is a characteristic of shopping lists that a vital requirement is sometimes forgotten! Although I have tried hard, and no doubt many others have too, I have been unable to find in the literature, or find for myself, any systematic method for drawing up an exhaustive list. Expressed in scientific terms there is no algorithm that enables one to draw up a comprehensive checklist for a specified part. So, while I am a great advocate of checklists, there is no guarantee that any list will cover all eventualities.

There is an additional difficulty: most designers regard checklists with scorn. An effective checklist must include the most obvious and trivial failure modes. Any good designer will have automatically considered these, and has no wish to be asked to do it again: indeed, he rebels against such imposition. One recognised way round this difficulty is to list only failure modes that are known to have been the source of field failures in the past. No designer can then say that they are irrelevant. After a long period of time a substantial checklist may be drawn up! But it cannot claim to be comprehensive.

So, while in principle checklists may be a powerful tool, the non-algorithmic nature of their construction implies that they themselves are unreliable. Furthermore, they are qualitative, and can make no contribution to any statistical interpretation. Checklists thus bear much the same relationship to design methodologies based on statistically defined inputs as they do to empirical design.

9.2 FMEAs and FTAs

As is well known, in executing an FMEA (**failure modes and effects analysis**) or an FMECA (**failure modes effects and criticality analysis**) one first identifies each failure mode at some designated level, be it component, sub-assembly, machine, or sub-subsystem level, and traces the effect of the failure through all the higher levels of the hierarchy in turn. In this way it is established whether each (i.e. ideally every) failure mode has unacceptable consequences on the system as a whole. If so, action can be proposed at an identifiable source in the lowest level. An FTA (**fault tree analysis**) is the reverse. It starts with an undesirable top event, and isolates possible causes at each successive lower level of the hierarchy in order to establish the prime cause or causes.

One common belief needs to be resolved at the outset. The acronym FMEA is always interpreted as 'failure modes and effects analysis'. The 'and' in this title is both unnecessary, misleading, and erroneous. An FMEA is *not* an analysis of failure modes; neither does it tell us what to do at the lowest level if the consequences are unacceptable. It is an analysis only of effects, which may or may not be quantified. The identification of each failure mode for input into the FMEA has to be carried out by other means, such as checklists, and corrective action is left to the ingenuity of the designer.

It is often stated, with some justification, that the FTA is the more powerful of the two techniques at the design stage, in that it forces the designer to consider all the causes of unacceptable top events. This is very true. The trouble is that the analysis is not pursued far enough, and the prime causes are not revealed. For example, an overheating bearing may be isolated as the source of an unacceptable event in a machine operating in a potentially explosive atmosphere. But there are a hundred and one causes for a bearing overheating, and the FTA as usually performed does nothing to resolve which, and hence identify appropriate design action. There is no fundamental reason why this should be so, of course, though there may be practical ones.

The weaknesses of both FMEAs and FTAs can be seen through their implementation by means of pro-formas or interactive computer programs. It follows that both techniques in their present form are algorithmic. Indeed, they were devised by the bureaucracies of large organisations (such as Defence) which live by algorithms. Consequently, the non-algorithmic stages of a complete analysis are not included; they cannot be, in most current implementations. Yet the non-algorithmic stages are all-important if the basic source of unacceptable behaviour is to be identified. So, neither FMEA nor FTA can make a direct contribution to the design activities that form the main subject-matter of this book. All they can do, if the top event is unacceptable, is call for a higher reliability at a basic level. In effect they are more an aid to specification revision than they are to actual design.

This does not mean that FMEAs and FTAs are of no use. They are probably more effective at the system, sub-, and sub-subsystem levels than for machines and parts. Even so, they do have a role to play in the latter scenario. Suppose, for example, an engine part can fail in a way that would sever a fuel line, with the possibility of a resulting disastrous fire; then one of these techniques would draw it to the attention of the designer – provided that the failure mechanism of the part at fault had been correctly recognised beforehand by other means. This could trigger immediate redesign action. Another and more general role, particularly in less extreme situations than the one depicted above, would regard the fire as a secondary failure, which would add to the cost of remedial maintenance action. These economic aspects of failure are pursued in more detail in the next chapter.

Kirrane (1989) describes three case studies in which these techniques are applied to mechanical sub-assemblies. They bring out very well the difficulties and the potentialities of the techniques. Anyone contemplating their use would be well advised to study these examples in depth before embarking on what could be a lengthy and expensive undertaking. Kirrane's conclusions are summarised below, with some additional observations of my own:

1. First he found that timing is critical. It must not be so early that insufficient detail is available to support the analysis, and not so late that the conclusions reached cannot be implemented. I would just observe that one of the most familiar expressions heard in any design office is 'We wouldn't do it like that if we could start again', or words to that effect. There never is time to start again or to make all the improvements that a designer would like to incorporate as experience of a project grows. Likewise for

FMEAs and FTAs: the window of opportunity is miniscule. Surely the analysis must be concurrent with the whole design activity, if it is to prove effective. Concurrency of all design activities (concurrent design) is now seen as essential.

2. Kirrane considers that the precise objectives should be defined in advance. I should have thought that this was the prime requirement of any undertaking, and not special to these techniques.

3. 'The boundaries of the system to be analysed, inputs and outputs across those boundaries, the build standard, the operating states and modes, any human errors (manufacture, maintenance, or operation) or other external factors, etc. to be considered should be defined in advance and not during the study.' Others have made the same observation in various phraseologies, so this conclusion has considerable support. But would not all designers like to work in such controlled conditions? It is just a fact of life that design is not only open-ended – it is wide open all round. It may be worth noting, though, that a design methodology based on statistically defined inputs has the same objective that Kirrane demands: accurately quantified design specifications.

4. Finally, Kirrane lays emphasis on communications, which must ensure that any conclusions from the analysis actually reach those who can implement them. I would fully endorse that in all circumstances.

As in the case of checklists, one reaches the conclusion that FMEAs and FTAs have no additional impact on the design methodology based on statistically defined inputs outlined in the previous chapters. They are of no use in resolving the imponderables: which distributions to use in the design, and so forth. They bear the same relationship to this kind of design as they do to current empirical design, and are of no greater nor of no less value. They do, however, have an important role to play in revealing the effects, and hence the costs, of failures. Numerical evaluation of costs can be more important in an overall design strategy when reliabilities are quantified, as will be discussed in section 10.2.

9.3 Design reviews and design audits

During any design the chief designer, or possibly the project manager, will inevitably hold many informal meetings with those actually carrying out the design, to correlate activities and monitor progress. However, it is highly desirable that occasional more formal meetings, known as **design reviews** or **milestone reviews**, be held, involving all those concerned with the design, both to review progress and to ensure adequate liaison. The major modus operandi of these reviews is through improved communication. Usually about three major reviews are considered adequate, though additional minor reviews may be necessary in a large project. A design audit is self-explanatory. It may be regarded as a super-review, often conducted at a somewhat later stage by people completely independent of the design team.

Now the real difficulty facing any designer is the definition of the input data to the design. The whole object of a statistically based design methodology is to facilitate this definition by expressing all input information quantitatively in well-established statistical terms. Once this has been done, the design can proceed on what is substantially conventional lines without the empiricism. But it is not the designer's job to second-guess what others are going to do, and the design review is a golden opportunity to resolve the many differences of opinion and undeclared inputs that surround much design. All concerned should be present. Many members of organisations using empirical design techniques, besides the designer, hide behind the factor of safety. They must be winkled out from the nooks and crannies in which they have taken refuge, and forced to declare their hands. For example, the load distribution to be used in the design must be provided, quantitatively, by the prospective user, the project sponsor, or the project manager, as appropriate. It must be agreed by all concerned in a fully documented form at an early design review. The source of all materials must be declared, together with their corresponding properties, expressed statistically. If these are not the 'as finished' properties, modifications introduced during manufacture must be declared by those responsible. The source of all bought-in parts must be declared and their reliability record provided. Any differences between prototype and production methods must be declared. The maintenance procedure, or at least an idealised version on which the design can be based, must be agreed by all concerned. Costings must be agreed. Reconciling and agreeing all such design input information falls firmly into the hands of the design review body, though active evaluation would be done outside any meeting, of course. The design review must ensure that it has been done, and that all concerned agree. If they do not agree, the design review is the arbiter. Because design is an iterative process, some of the data will inevitably have to be changed, but when that is done, changes must be agreed at a subsequent design reviews: i.e. when all those concerned are present, or at least have access to, and hence knowledge of, the changes.

Most of this work would not be carried out by a design review centred on an empirical design process. In that methodology most of the above aspects would be concealed under the empirical factor of safety, which no one wants to discuss. The design review therefore takes on an additional role when a design methodology based on statistically defined input data is being implemented. Indeed, reviews have a key role to play in these circumstances, which must be fully comprehended by the chairman and secretary of a design review committee or body. All this must also be taken on board by the design audit, which has to endorse, or at least take note of, all the statistically defined inputs to the design. So, while the part played by checklists, FMEAs and FTAs is little different in empirical design and design based on statistically defined inputs, design reviews and design audits should be an integral part of the latter methodology, with a much more significant and important role to play.

References

Carter, A.D.S. (1986) *Mechanical Reliability*, 2nd edn. Macmillan, Basingstoke.

Kirrane, P. (1989) Applications and problems with the use of mechanical design evaluation techniques, in *Seminar on Improving the Reliability of Mechanical Systems at the Design Stage*, Institution of Mechanical Engineers, London, pp. 17–21. Also in *The Reliability of Mechanical Systems*, Editor J. Davidson, Technical Editor 2nd edn C. Hunsley, Mechanical Engineering Publications, London (1994), pp. 140–145.

10
Design strategy

10.1 The overall design activity

So far in this book we have been concerned with the methodology of design: i.e. with what may be described as its tactical aspects. It is now necessary to put these into the context of an overall design activity, i.e. to consider design strategy.

There is one thing that 'sticks out like a sore thumb' in any non-epiphenomenological statistical investigation of reliability, and indeed is a recurring feature of many preceding chapters of this book. It is sensitivity – made more acute by the indeterminate nature of load and strength distributions. Even if we were to take the most optimistic view, and assume that rigorous quality and process control will eventually solve all the problems associated with strength and damage resistance distributions, the load distribution is still often an enigma, for the reasons already discussed. I find it quite amazing that even some reliability technologists still regard reliability as an inherent property of a part, component, or system, which, for example, can be recorded in data banks, substantially ignoring (other than maybe by some crude factoring technique) the load to which it is subjected. A product is not of itself 'reliable' or 'unreliable', but only so in the context of a loading environment. Because this can differ so much from application to application, the best that an *ab initio* design can hope to achieve is a hazard or breakdown rate that is within, say, an order of magnitude (perhaps several orders of magnitude) of any specified value. It may well be possible by over-design to achieve such low hazards or breakdown rates that an order of magnitude or whatever is of no concern, but that is not optimum first-class design. How then is high reliability, compatible with the best performance, to be achieved?

I would contend that there is only one situation in which the effective load distribution can be discerned, and that is in the hands of the user: i.e. in service in the field. Simulations for prototype testing have to be made, but they are often only pale imitations of the environment that will eventually be encountered. If therefore we wish to know how a product will behave in

terms of reliability we must turn to actual use. It is thus good practice to start
with an existing product, and to record precisely how it is behaving in the
hands of actual users. This may not usually be considered where design be-
gins, though it is a common scenario for the start of reliability programmes,
and must be for design. Every failure must be recorded, as and when it occurs in the field. Records must be as comprehensive as possible, must be
fully analysed, the cause or causes of failure established, and the design assumptions that led to the failure fully revealed. It may be worth adding that
if, say, a part failed because of a surprisingly high load, then it is not good
enough to note this as the cause of failure and dismiss it as 'exceptional'. If
it happened once, it could happen again – at some probability. All incidents,
including exceptions, must be given a probability, to be used in future designs. It seems unnecessary to add that corrective action must follow, but so
often action is obviously not taken that one must do so. There can be no
exceptions to this action either. If appropriate action is always taken, an acceptable product will eventually be achieved.

To illustrate what can be achieved by this policy, we turn to aero-engines
– admittedly a field where reliability is a prime requirement. Figure 10.1 shows
a plot of civil aero-engine in-flight shut-down (IFSD) rates plotted against
cumulative usage. It is CAA data published by Vinal (1991). The continued
reduction of IFSDs leading to the current very low values for contemporary
aero-engines is immediately apparent. However, while this plot is admirable
for its intended purpose – the justification of the oceanic twin jet – it does
not tell us how it was achieved. As a step towards that objective, the data
have been replotted using Duane axes in Figure 10.2. (For the reader who is
not familiar with Duane, or the more advanced AMSAA analysis, the author
has given a simple description (Carter, 1986) with further references to more
detailed work. However, the immediately relevant features will be described
in the subsequent text as required.) In Figure 10.2 each plotted data point is
the mean of the range in Figure 10.1 for each cumulative flight time. In com-
pliance with Duane procedures, logarithmic scales have been used for both
axes. According to Duane (1964), such data should plot as a straight line
during any development programme. In producing Figure 10.2 Duane's ideas,
which were derived for single-product development, have been expanded and
applied to generic data over an extended period of time. The data again follow a straight line.

Duane's equation for his straight line is

$$\lambda_c \, (\Sigma t) = \frac{K}{(\Sigma t)^v} \qquad (10.1)$$

where $\lambda_c(\Sigma t)$ is the cumulative failure rate at time Σt (in Duane terminology). In the terminology of this book cumulative failure rate should be replaced by cumulative breakdown rate, $\Sigma W/\Sigma t$. Σt is the total or cumulative
operating time (of all items in the population), ΣW is the total number of

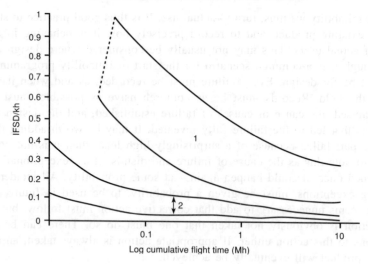

Figure 10.1 Total in-flight engine shut-downs versus cumulative flight time: 1, limits for early turbo-jets and turbo-fans entering service before 1981; 2, limits for derivative and new technology engines entering service after 1981. Data from Civil Aviation Authority, published by Vinal (1991).

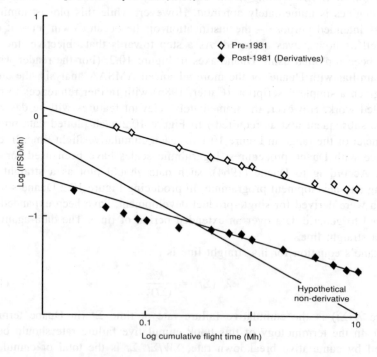

Figure 10.2 Duane representation of data in Figure 10.1.

breakdowns (experienced by all items in the population), and v, K are constants.

Expressing equation (10.1) logarithmically:

$$\log [\lambda_c(\Sigma t)] = \log K - v \log \Sigma t \qquad (10.2)$$

giving a straight-line plot on log–log scales as in Figure 10.2. Note that the data plotted in Figure 10.2 are actually the instantaneous breakdown rate $\lambda(\Sigma t)$ quoted by CAA, but the cumulative curve will be a straight line parallel to that plotted. The instantaneous rate is derived from the cumulative rate using

$$\log [\lambda(\Sigma t)] = \log [\lambda_c(\Sigma t)] + \log (1 - v) \qquad (10.3)$$

so that both rates plot as straight lines on logarithmic axes.

The Duane index, v, measures the intensity of the reliability growth. It is the slope of the line on the log–log plots. It is usually quoted as having a value of 0.4–0.5 in a normal prototype development programme, rising to 0.6 in one that concentrates on improving reliability growth. The value for both the lines drawn in Figure 10.2 is 0.3. The CAA data obviously do not include pre-service prototype development carried out by the manufacturer; neither do they include the first 10^4 hours in service operation, when the early learning phase would give rise to some reliability growth not related to design. The growth in Figure 10.2 has thus been achieved during normal operational use. It can only have been achieved by correcting all sources of in-flight shut-down as they were revealed. The fascinating feature is, however, that the same value of the index has been maintained over the last 20–30 years, being no better now than 30 years ago (some consolation to us older engineers!). But, being maintained over that time span, it has resulted in the modern highly reliable aero-engine. Actually, the surprising feature is not so much that the index is no better now than it was 30 years ago, but that it has been possible to hold it at that value for such a long time. One of the early criticisms of the Duane model was that reliability growth would reach a limit that was not modelled by equation (10.1). Consequently, several modifications to the Duane equation were suggested. They would seem to be unnecessary.

Not all industries can be expected to achieve the high reliability growth of the very reliability-conscious aero-engine business, of course, but it is suggested that any industry that has reliability as one of its selling points should aim at a sustained Duane index of about 0.2 (or perhaps slightly higher). If the aero-engine industry, which starts with a highly reliable product, can sustain an index of 0.3 over 30 years, the lower value is certainly possible with run-of-the-mill products. Even so, it is not easy. Not every industry has a CAA to collect and analyse data on a systematic basis. Without data it is impossible even to start. Designers must ensure that resources are available to collect the data, and that the data reach them. Even when this is done, designers need to check the data thoroughly, for the author has found that nearly all data collection and reporting systems are corrupt to some extent. For example, product

support groups or departments can block or distort much useful information
when it is obtained. They specialise in the 'quick fix' or 'band aid' solution,
which provides the short-term solution so attractive to the user and the manufacturer's financial director alike – sometimes referred to as the 'white knight'
approach – whereas what is essentially required is an in-depth study of the
design assumptions that led to the failure. Finally, if a proper solution is
found, managing directors and production managers are reluctant to implement it – usually on the grounds that it is too disruptive of an established
production line, or too costly. The introduction of NC machines and robotics
could remove many of these objections, in that changes are easily made and
no established human practices have to be overcome. Nevertheless, there are
at present significant obstacles to be surmounted if an organisation is to follow the policy advocated to achieve an acceptable Duane index. The responsibility must rest with the designer. For, sooner or later, a new product will be
required, and data for a new design can come only from a proper understanding of past field failures.

In the case of aero-engines, such a new demand came in the mid-1960s.
For performance reasons, which need not concern us here, the civil market
demanded much higher bypass ratio engines, though the military market still
concentrated on the lower bypass reheated engine. The resulting experience
of the late 1960s and early 1970s was salutary. Prior to this, failures had
steadily declined, as shown by the upper curve in Figure 10.2, but the lessons
clearly did not get through to the designers. Many of the new high bypass
ratio designs were not initially reliable, and consequently the development
costs to achieve the targets proved excessive. Of more importance, it was
soon noted that the more innovative designs were the least reliable. By the
1980s the concept of the derivative engine had been born. As its name implies, a derivative engine is developed from an established proven engine
with as few major changes as possible – preferably only one. The idea was
by no means a new one, but the concept of derivatives was formulated as
established policy at that time. Thus design became a continuous evolutionary process, with each step fully proved by field experience before the next
step was taken. I would suggest that the term 'new technology' used in Figure 10.1 is an emotive one, of no particular relevance in connection with
reliability; indeed, it is a misleading one. The technological leap from piston
engines to gas turbines was a far greater technological step than from low to
high bypass ratio turbines. Continuous evolution is the underlying reason for
the in-flight shut-down pattern shown in Figure 10.2. It seems to me the only
way to deal with the intractable environments that undermine all theory. It
may be argued that this is only old-fashioned development, rigorously applied. Why bother with improved design methodology? Why not circumvent
the difficulties of design by 'doing as best one can' as quickly as possible
with existing wholly empirical methods and let development look after reliability – i.e. do what we have been doing since the beginning of the cen-

tury? I would contend that this approach is now too expensive.

To examine this contention more closely, let us suppose that a manufacturer of aero-engines sets out at the time of change from low to high bypass ratio engines to make a non-derivative (i.e. a totally new) engine to compete with the derivatives in Figure 10.2. In effect he is returning to the beginning of the non-derivative curve in Figure 10.2. However, it will be assumed that, because technology would have advanced, he can do somewhat better than this: half way, say, between the non-derivative and derivative curves, after prototype development and 10 000 hours' in-service experience. His starting point is shown in Figure 10.2. Actually this assumption is optimistic. By labelling one curve in Figure 10.1 'derivative' CAA (or Vinal) implies that the other curve is non-derivative. In fact, most of the engines to which that curve refers are also derivatives – derivatives of military engines. The complete history of gas turbine aero-engine reliability would have a third curve above the two shown, but the data are all shrouded in military security and not available. The original data published by Duane suggest that the log of the cumulative IFSD rate at 10 000 cumulative flight-hours is about 0.45, which gives some idea where the line would lie, though Duane's (1976) data refer to the early in-service history of a single engine type. From his personal experience, the author can assure the reader that very large sums of money indeed were spent on achieving acceptable reliability of military gas turbine engines in the early days – paid for by 'cost plus' contracts funded by the taxpayer.

In spite of these reservations, let us suppose that our hypothetical manufacturer can achieve the target shown at 10 000 cumulative hours in Figure 10.2. The manufacturer appreciates that the reliability of his new engine will lag behind that of his competitors' derivatives, though of course he hopes to achieve a much better performance. That is the reason for his proposed design. So he sets up a more intensive extended development programme to achieve greater initial in-service reliability growth. We shall be optimistic, and assume that he achieves what is usually taken to be the maximum normal Duane reliability growth index of 0.6, giving the line shown in Figure 10.2 for the expected behaviour of his non-derivative engine. It will be seen that it intersects the line for generic derivative engines at 270 000 cumulative hours: i.e. this is the reliability break-even running time. But think what this means. Using ten prototypes running 24 hours a day 7 days a week, for 52 weeks a year (impossible to achieve by a factor of maybe 5) would require a development programme of 3 years to make the non-derivative engine's reliability the same as that of the derivatives' – 15 years if the factor of 5 applies! The alternative is to provide a much larger number of prototypes and test facilities. A normal firm could just not finance such a programme, and the reliability of the new engine would trail behind that of its competitors with all its consequences. This example seems to me to back up the general conclusions to be reached from Duane plots: proceeding with very innovative designs, which cannot be supported by an accurate design methodology, and relying on subsequent development to overcome any design deficiency, is inviting disaster.

The difficulty is the time scale involved. In the case of performance the problem does not arise. For example, if one were trying to outdo the competition by achieving a better specific fuel consumption, any design innovation could be verified immediately on the testbed. The laws of thermodynamics being practically deterministic (though fundamentally statistical!), the fuel consumption measured on prototype test would be valid for the whole life, give or take a very small acceptable deterioration. To measure reliability, on the other hand, requires testing over a long period of time: indeed, a significant proportion of the lifespan itself, during which the competitors are advancing the target at almost the same rate. In the reliability field one has to run to stand still. It bodes ill for the innovator (unless he has a substantial performance improvement on offer), or indeed the producer who has fallen behind. As an example of the latter, we need only turn to the motor car industry. For reasons that again need not concern us here, after World War 2 the reliability of UK cars fell significantly below that of Japanese-designed and manufactured cars, indeed well below prewar UK standards. The customer responded appropriately, with well-known results. In the last decade or so, UK companies have made a determined effort to improve the reliability of their product, with considerable success; but the Consumers Association (1994) still reports UK car reliability below Japanese. People who ask why should consult Duane charts. The lesson is that it is extremely difficult to catch up a reliability deficiency in a long-established complex product by development testing or 'experimental design'.

Prototype development can be only a fine-tuning process for a substantially acceptable design, so far as reliability is concerned. It can, as noted above, be of more direct value in achieving performance targets that can be directly and uniquely measured on the testbed, but that is not the subject matter of this book, and the two objectives should not be confused. Any organisation must plan its design and development programme to meet these constraints. Unless he can be sure of adequate resourcing should he fail to achieve his reliability target (or the risk is a calculated one considered worth taking), a designer would do well to adopt a circumspect attitude towards innovative design for reliability – which leads us back to the beginning of this chapter.

The most accurate (and cheapest) testbed is the user. Field experience must be the source of design data. It is the reason why field experience must be fed back directly to the design team. The maximum amount of generalised design information has to be extracted from these data. For example, if a part has failed, it is not good enough to eliminate the failure by making it a bit stronger, i.e. by increasing the factor of safety in the usual arbitrary way. It is necessary to know whether all the material properties (their mean, standard deviation, shape of distribution etc.) conformed to the design assumptions. If not, why not? Even 'exceptions' must come from some distribution. It is necessary to know whether the load (its mean, standard deviation, shape of distribution etc.) conformed to the design assumptions. If not, why not? Were some loading conditions not considered at all? If not, why not? Otherwise, why did the part fail? In all this, quantification is the essence of success. That is why the statistical representation of loads and strengths is so important. The information is not only necessary to deal immediately with failures. The same information is necessary to assess the design and subsequent redesign of all parts: those that have not failed, as well as those that did. Non-failures themselves tell us virtually nothing.

For obvious reasons, the aero-engine industry is very reliability conscious, and so lies at one extreme end of the reliability spectrum. If the Duane generic long-term reliability growth index for some other product were found to be much less than that achieved with aero-engines, a concentrated development programme at an index of 0.6 would look a much more attractive proposition. Against that, an organisation that can sustain an index of only 0.1, say, in normal usage is not going to achieve an index of 0.6 in a concentrated programme. The resources in personnel and equipment will just not be there. Each case is special. Each organisation must plot its own version of Figure 10.2. It is very simple, as initially the advanced techniques of the AMSAA model are not required. When that is done the optimum strategy can be defined, but nearly always it will be found that development is a very expensive route to reliability. The key to success is a close approximation to the final product at the design stage. That is the justification for the earlier chapters of this book, and for the attention that has been paid to methodology.

10.2 Design for economic reliability

Reference has already been made in section 7.2, apropos of sensitivity, to the economic consequences of the extended life distribution. In this section a closer look is taken at the economics. It applies in force to the extended life distribution but also, though perhaps with less restrictive consequences, to more compact distributions and hence to all design. Before proceeding, the reader should be warned that much of this section could be regarded as controversial, though taking that stance does not get rid of the problem!

It will be recalled that although finite standard deviations of the damage resistance are the source of life distributions, the main reason for the more extended distribution is the low slope of the s–N curve. It is therefore a design rather than a quality problem. (Knees or roller-coasters are a quality problem, and are assumed to have been already eliminated.) Unfortunately, in many cases there is little that the designer can do to curtail these distributions, because the slope of the s–N curve is an inherent property of the material that he uses. Choice of material and manufacturing techniques can sometimes mitigate the situation, but usually only marginally so. Design decisions have to be based on other criteria.

Possible design action can best be illustrated by means of a simple example. Suppose we wish to design a straightforward tension member, which is to be part of a complex machine. It will be supposed that the load to which it is subjected is Normally distributed with a standard deviation 30 per cent of the mean: a not untypical case. It will also be supposed that this load is to be applied 10^{10} times during the planned lifetime of the part, i.e. before the parent equipment is retired or 'cast'. The part is to be made from a material whose fatigue limit is Normally distributed with a standard deviation of 7.5 per cent of its mean. Actually, a Weibull distribution with a shaping parameter of 3.44 was used in the calculations that follow. This was done to provide a defined lower cut-off to the threshold distribution, thus avoiding having to make arbitrary cut-offs, as is necessary when using the Normal distribution itself. The fatigue limit is reached at 10^6 cycles. The s–N curve itself is linear on semi-logarithmic axes between the fatigue limit at 10^6 cycles and the UTS at 10^3 cycles. The UTS is twice the fatigue limit. These values have been chosen to be typical without an unduly emphasised low slope to the s–N curve. The extended life effect is thus in no way exaggerated by these assumptions. No reliability will be specified. If the customer does specify a reliability at a given life, there is nothing further that the designer can do, because the design is fully specified, and the customer is always right! Instead, for the purpose of this illustration, a series of designs were carried out using a range of damage margins. The highest was chosen so that no physical fatigue damage was done: i.e. the design would be intrinsically reliable with respect to the fatigue limit. This is a common design criterion. Any planned life is irrelevant to this design, because no physical damage and hence no fatigue failures could occur no matter how long the part was operated. The next (lower) damage margin was chosen so that although physical damage would be done, no failures would occur in the planned life. Failures would arise, of course, if operation were continued beyond 10^{10} cycles. Then a number of damage margins below this were arbitrarily selected. With these margins, failures would arise before 10^{10} cycles. A repair maintenance policy was adopted to take care of these failures: i.e. the part would be replaced on failure without any prior scheduled replacements. This is the optimum policy with a falling hazard, such as would arise in this high-cycle fatigue process. The necessary replacements were calculated using Hastings's program (1967). In order to assess the cost involved it was assumed that the initial cost was proportional to the weight, i.e. to the cross-sectional area of the part; its length was constant for functional reasons. This is thought to be a very reasonable assumption for a series of geometrically similar parts made in the same material. It was then assumed that a replacement part would cost 2.5 times the cost of an initial part. This factor was introduced to allow for any dismantling of the machine during repair, and for any revenue lost while the repair was being carried out. It is also supposed to include other less tangible costs, such as loss of goodwill on behalf of the customer etc.

For an assumed value of the damage margin, DM

$$\frac{\bar{E} - \bar{L}}{\sqrt{(\sigma_E^2 + \sigma_L^2)}} = DM \tag{10.4}$$

or

$$\bar{E} = DM \sqrt{(\sigma_E^2 + \sigma_L^2)} + \bar{L} \tag{10.5}$$

$$= DM \left(\frac{\sigma_L}{LR}\right) + \bar{L} \tag{10.6}$$

Substituting for $\sigma_L = 0.3 \bar{L}$:

$$\bar{E} = \left(\frac{0.3DM}{LR} + 1\right) \bar{L} \tag{10.7}$$

or

$$\bar{E} = \left(\frac{0.3DM}{LR} + 1\right) \left(\frac{\bar{L}(p)}{u}\right) \tag{10.8}$$

where u is the cross-sectional area and $\bar{L}(p)$ is the mean load (force) applied to the part. Let c be the initial cost per unit cross-section. Then

$$\text{Initial cost} = c\, u = \left(\frac{0.3DM}{LR} + 1\right) \left(\frac{c\bar{L}(p)}{\bar{E}}\right) \tag{10.9}$$

where c, $\bar{L}(p)$ and $\bar{E}$ are constant for all values of the damage margin and its associated loading roughness, LR. The costs expressed in units of $c\bar{L}(p)/\bar{E}$ are given in Table 10.1, assuming a population of 100 parts. In the table the total of surviving and replacement parts is sometimes greater than the initial population of 100. This occurs when some of the replacement items have themselves failed.

Table 10.1

Damage margin	2.75	3.0	3.5	4.0	4.6*	5.4+
Reliability at 10^{10} cycles (%)	19.5	42.5	78.5	99.0	100	100
Survivors	20	45	80	99	100	100
Unscheduled replacements	240	102	22	1	0	0
Initial cost	1.92	2.00	2.20	2.40	2.66	3.03
Replacement cost	11.52	5.10	1.21	0.06	0	0
Total cost	13.44	7.10	3.41	2.46	2.66	3.03

*no failures
+no physical damage

The above non-dimensionalised total cost has also been plotted against the damage margin as curve (a) in Figure 10.3. Significantly, the total cost has a minimum value, and the minimum value corresponds to a particular reliability. Some find this surprising. But Figure 10.3 is no more than another version of the diagram in Figure 10.4, which is always regarded as self-evident, is fully accepted, and appears in numerous textbooks on reliability and endlessly at conferences on the subject. The reliability plotted in Figure 10.4 can be controlled, so far as the designer is concerned, only by choice of damage margin. It is damage margin that controls the interference of the relevant distributions and hence the reliability. There is therefore a direct relationship between the abscissae on Figures 10.3 and 10.4. Of course, the designer has a number of ways in which he can effect a change of damage margin, each with its associated cost penalties, but the basic shapes of the curves in Figures 10.3 and 10.4 must be the same.

Looking at these curves more closely, it will be appreciated that the rising arm of the curve for damage margins greater than the minimum is due to the increasing initial cost, and the rising arm for damage margins less than the minimum is due to increasing maintenance costs. The minimum is a balance of the two trends. For the tension member, the minimum cost corresponds to a reliability of 99 per cent, even though a reliability of 100 per cent at the chosen life is possible: i.e. the optimum design based on cost implies deliberately designing for failures! To put this further into perspective, it will be seen that the cost of a part to avoid all damage – a not unusual design criterion – will be some 23 per cent in excess of the minimum cost. It is not the kind of figure that can be neglected. The customer certainly will not ignore it. Also, note that if minimum cost is the design criterion, one is not at liberty to choose or specify a reliability. The reliability is determined by the individual costs that make up the total cost. These individual costs must therefore be accurately evaluated, but this is not easy.

In the first place, increasing the weight of a basic part must involve further increases in the weight of other parts: supporting parts that have to carry the additional weight, for example, in matching parts, and so on. It follows that there is a 'knock-on' effect of any weight increase – or of any weight decrease. In designs falling into the author's experience, it was often found that the total weight increase of a reasonably complex machine was about four times any basic part weight increase. There is no reason to believe that this figure has any universality (in some cases no increase may be involved), but using it as illustration gives the curve (b) in Figure 10.3. The damage margin for minimum reliability is reduced from 4.0 to about 3.8. There is, however, a more serious effect, in that any weight increase will reduce the payload in some way. It is more serious in that it reduces the earnings or revenue from the machine, or perhaps many machines, over the complete lifespan. It is not a one-off cost. For example, increasing the weight of every compressor blade in an aero-engine by a small amount could lead to the loss of one fare-paying

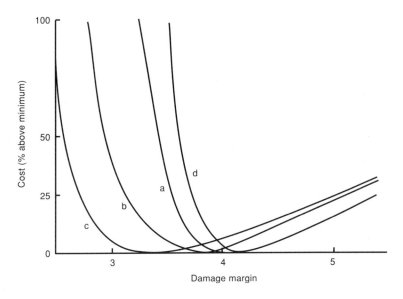

Figure 10.3   Cost of tensile member; a, b, c and d defined in text.

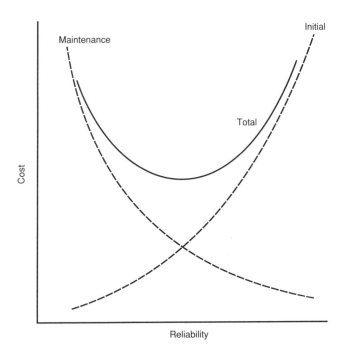

Figure 10.4   Standard curve of cost versus reliability.

passenger place in a four-engined aircraft. Taken over the lifespan of the aircraft, the loss of revenue from a fleet would be high. The penalties of over-weight design can be very high indeed; but conversely so can be the rewards for weight reduction by good design. The cost penalty due to loss of revenue will vary significantly from one application to another. Purely as an illustration, it was conservatively assumed that this cost was equivalent to a further knock-on effect of ten times the initial cost for the simple tension member, to give curve (c) in Figure 10.3. This reduces the damage margin for minimum cost to about 3.35, which is a very significant decrease. The corresponding reliability at the planned life is about 75 per cent: well below 100 per cent, and not the kind of figure that is usually bandied about! Anyone who thinks that this figure is absurd is referred back to the data in Figure 7.7 for the automotive engine water pump. Its full life reliability was about 70 per cent at a vehicle rated life of 60 000 miles, falling to 55 per cent at 100 000 miles. These reliabilities are even lower than those calculated for the tensile member! These are figures for a well-developed pump, which was considered quite adequate for the job.

As already pointed out, increasing total costs can also arise from increasing maintenance cost associated with more failures at reduced margins. Maintenance costs here include not only the remedial cost associated with the primary failure, but also that associated with secondary damage, which in extreme cases can be many times that of the primary failure. FEMAs and FTAs have an important role to play in identifying these consequential costs. The cost of the active maintenance is not difficult to establish, but other costs are involved: for example, the loss of revenue while the machine is out of action; but this is more a function of the time waiting for a replacement and perhaps for a repair team, than of the time for active repair. These in turn depend on administrative arrangements, which vary considerably from one organisation to another. It is a non-technical aspect that is difficult to evaluate. Worse still is the loss of goodwill on behalf of the customer, who could take his next order elsewhere if annoyed by a (or a series of) unexpected breakdown(s). But his purchase could be a one-off, so there then will not be repeat orders anyway! Goodwill is almost impossible to assess in general terms, but allowance has to be made for all these aspects when estimating the cost of a failure in real life. Curve (d) in Figure 10.3 shows how the damage margin for minimum cost is increased when the cost of a replacement is increased to 10 times the initial cost. Qualitatively, it is very easy to see what must be done, but quantitatively it is very much more difficult to say by how much.

One of the significant features brought out by Figure 10.3 is the rapid increase in total cost due to an excessive maintenance requirement if too low a damage margin is used in design. It pays to beware, if costs are not accurately known.

As a more specific and real illustration of these problems, let us return to the water pump of Figure 7.7, setting the design life equal to a vehicle casting life of 100 000 miles. The Weibull distribution shows a failure-free life of 12 000 miles. Taking the average vehicle distance covered per year as 10 000 miles, the manufacturer could offer a one-year free replacement warranty, with the probability of having actually to make very few replacements. Assuming that all the vehicles cover a full life of 100 000 miles, at which the pump reliability is 55 per cent, roughly half the pumps will have survived and half will have been replaced. This seems a reasonable 50/50 split between supplier and customer, without getting involved in any intractable cost calculations. The customer was apparently satisfied and the manufacturer, having made a sizeable additional profit from replacements, is also happy. However, using the above Weibull distribution it can be shown that we must expect about 14 per cent of the replacement pumps also to have failed by 100 000 miles. I cannot answer for the reader, but I would consider myself merely unfortunate in having to replace a water pump once in my car, but would consider it a 'bit of a rip off' if I had to do it a second time. By contrast, someone concerned with a large fleet of vehicles may be quite happy with this state of affairs. Some will be replaced once, some twice, but some not at all – 'you win some and lose some'. It is interesting to note that if the manufacturer considered 14 per cent second-generation failures excessive, and increased the failure-free life to 30 000 miles to give no second-generation failures before 60 000 miles (a good psychological point), then one would expect 44 per cent failures at 100 000 miles. This would reduce the manufacturer's profits from the sale of spares by just over 15 per cent. Presumably he would have to recoup this loss by increasing the price. Which would the customer prefer? It is not at all clear, and is as much as a psychological matter as it is technical. The reliability of the improved pump at 100 000 miles would still only be just over 60 per cent.

The pump reliability at the planned life of the parent equipment just quoted should not be confused with the reliability of a pump, which is currently functional, completing a given task or mission: i.e. the probability of operating successfully for a smallish fraction of the total life. The pump's mission reliability for a mission of 1000 miles, i.e. the probability of the vehicle completing a further 1000 miles without a pump failure, is plotted against vehicle age (distance covered since new) in Figure 10.5 for the original design of pump installed in a vehicle and only replaced on failure. It seems just acceptable to me, though the reader may have higher expectations. He or she might also like to consider the consequences of specifying a 1000 mile mission reliability of 99 per cent; the author's observations on such a specification are given in Appendix 2 to this chapter.

The original part used to introduce the extended distribution, the tension member, is a very simplistic one, chosen only because parameter values can be specified easily and the calculation followed through to a firm conclusion. In the practical case the weight of the end fixings would probably dominate that particular problem anyway. The cost input data for a real design would,

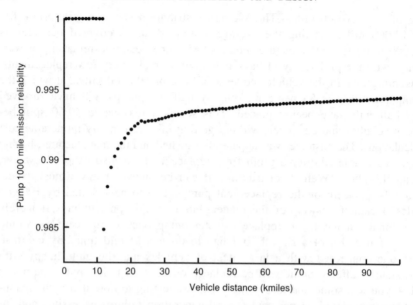

Figure 10.5 Reliability of water pump for a 1000 mile mission plotted against
parent vehicle age (× 1000 miles). There is a sudden loss of
reliability when all pumps reach the end of their failure-free
life together. This is not repeated (although a slight drop is
apparent at 22 000 miles) owing to randomisation of the repair
process. The reliability continues to increase with vehicle age
owing to replacement of short-lived pumps by long-lived ones
– survival of the fittest.

of course, have to be based on the practices followed by the designer's own
organisation, though the designer must ensure that it is the comprehensive
total (i.e. whole life) cost. But the principle is universal. While for some poor
and incompetently executed designs it may be possible to achieve higher reliability at no extra cost – or even at reduced cost – for competent designs in
any sophisticated application higher reliability means higher initial cost and,
of course, lower maintenance costs. An optimum therefore exists. The optimum reliability can often be well below 100 per cent. In many circumstances
there would then be little difficulty in measuring the required material damage resistance. The difficulty of measuring the material properties is replaced
by one of estimating costs.

   At numerous lectures I have given on this topic over the past years, there
has always been someone present, or so it seems, concerned with safety. They
all object very strongly indeed to design for reliability other than 100 per
cent – or perhaps more truthfully to any other than one for which a reliability
of 100 per cent cannot be claimed. They apparently wish to continue using
factors of safety, which they claim achieves this objective; at least, they do
not want to get involved with statistics. I cannot agree. The cost of an injury

or death is, of course, very high indeed and must, according to the ideas being advanced in this section, be debited to the cost of any relevant failure. This would force the optimum to. high damage margins in an exaggerated form of curve (d) in Figure 10.3. This could place the design in a sensitive region, in which case design is best based on a no-damage criterion using stress-rupture methodologies. The design would then be, or at least could genuinely be claimed to be, 100 per cent reliable. I cannot therefore see the problem, other than the universal one of defining the load distribution. From post-lecture discussions I have formed the opinion that those concerned with safety often do not want to quantify the basis of their designs. Liability in the event of failure seems to be the underlying motivation. The factor of safety certainly provides an excellent smokescreen. Emotion plays a very significant role in safety, and if that – or any other design parameter – cannot be quantified, statistical design is of little use. I recognise that designers have to operate in the ambience that surrounds the application of their products and act accordingly, but surely they should also operate with technical integrity.

Before leaving the subject of safety, it should be emphasised that the reliabilities being quoted above refer to potential failures. If potential failures can be prevented from becoming real failures by on-condition maintenance or fail-safe techniques etc., then the situation is different. In these circumstances the overall reliability depends as much on the reliability of the maintenance procedures or fail-safe devices as on the reliability of the part (see equation (8.1)), though the total cost does not. Target reliabilities have to be assessed using the whole environment.

The conclusion to be reached in this section is that design target reliabilities should be set to achieve minimum cost – recognising that particular difficulties may arise when safety is involved – and that, for some extended life distributions, minimum cost may result in reliabilities well below 100 per cent.

Even so, I have a slightly uneasy conscience. The cost to whom? No problem for the mass-purchaser, but the purchaser of a single item (a consumer durable – or perhaps a nuclear reactor!) may receive an item with a lifespan 100 times better than average or a lifespan 100 times worse than average. And both are charged the same! Of course, one would not expect all the parts of a given machine to be short-lived, but clusters of short-lived items are not all that exceptional: otherwise we should not have common expressions for them – such as 'the Friday afternoon car' – though I doubt very much if they are made on a Friday afternoon. In my youth the expression was 'it has a jinx on it', so the phenomenon is not new. There must also be clusters of long-lived items, of course, but they do not generate complaints! Individual purchasers are clearly not being treated equitably, but I see no solution to this problem.

There is a further problem that was noted, but not pursued, earlier. It concerns the cost of collecting data to facilitate a statistical design. It has been

demonstrated above that the desirable objective is low total cost – not necessarily high reliability. If, therefore, a simple low-cost, on-condition maintenance technique is available to prevent actual failures, design methodologies
based on statistically defined inputs can themselves become uneconomic and
unwarranted, because of the cost of collecting data. Each situation has to be
carefully assessed. If quantification is required, I would suggest that consideration be given to assessments at lower confidence levels. There is little purpose in working at a high technical confidence level if it cannot be matched
by the often equally important non-technical aspects.

The conclusion to be reached in this section, then, is that higher and higher
reliabilities are not necessarily the prime objective; lower and lower total
cost nearly always is. Reliabilities and design techniques must be chosen to
achieve the desired objective.

10.3 Reliability apportionment

Reliability apportionment usually refers to the allocation of target reliabilities
to subsystems in such a way that the specified reliability of the system is
achieved. Apportionment is obviously of considerable importance when the
subsystems are being supplied by different contractors, and each has to be
allocated a reliability. The procedure is well documented in the literature.
This section deals with the same problem but at a lower level; if the reliability of a machine has been specified, on what reliabilities must the design of
the constituent parts be based? It is again a critical problem. It is the reliability of the complete machine that will be specified, but the design methodology discussed in the previous chapters is based on a known reliability for
the constituent parts. The two have to be matched. This problem is not addressed in the literature.

Subsystem reliabilities are usually assessed on the basis of past experience
(data banks etc.) and coordinated via reliability block diagrams on a trial-
and-error basis to give the required result. This is nearly always based on
constant breakdown rates. When dealing with machines and parts this assumption cannot be made. The assessment must use the real behaviour of the
parts. There are also usually far more parts in a machine than subsystems in
a system, so the number of alternatives is much greater. Trial and error is
then no longer a practical proposition.

If the part under design is subject to a stress-rupture failure mechanism, no
problem arises. It must be designed for intrinsic reliability: i.e. effectively
100 per cent reliability for the machine's whole life (and in fact beyond). If
it is subject to scheduled replacement, then no difficulty should arise in allo-
cating a design reliability to the part. It is simply the T_F for economic replacement. If, however, the part under design is expected to remain in the

machine for its full life, being replaced only on failure, then difficulties do arise. One could of course choose the reliability of all the parts to be equal, and commensurate with the overall target. But that is very restrictive. Examination of contemporary designs shows that the parts usually have an increasing T_F (for a given F) for increasing complexity, cost, and maintenance requirements of the parts: i.e. the reliability of the parts is graded. Thus in an ordinary internal combustion engine one would expect the block to be much more reliable than, say, a cooling water hose fitting. There is nothing wrong in principle with this; in fact it is an eminently sensible procedure, but some criterion is necessary to allocate (apportion) the T_Fs.

Suppose the mean cumulative maintenance cost of a machine is given by the oft-quoted expression

$$\sum \bar{c} = \text{constant} \times t^{\omega} \tag{10.11}$$

where $\sum \bar{c}$ is the mean cumulative maintenance cost for the whole population at time t, and ω is an index that increases with the complexity of the machine: as a guide one would expect values of 1.3 to 1.6 for automobiles and about 2.2 for tanks and locomotives. Then it can readily be shown – see for example Carter (1986) – that the average mean total cost per unit operational time reaches a minimum at some value of t. The total cost is here the sum of the initial, maintenance, and operating costs. Obviously, it is not necessary for equation (10.11) to be an exact model for this situation to arise. Equally obviously, the machine should be operated up to this time of minimum total cost, because the average cost per unit time will be continuously decreasing from introduction into service until that time is reached; but the machine should then be withdrawn from service because the average cost per unit time will increase indefinitely thereafter. This minimum cost life (durability) should therefore be made the design life of the machine. To make this minimum cost itself as small as possible, it must also be the life for minimum cost (as discussed in the previous section) of all the constituent parts. The rule for apportioning parts reliability is therefore that their life for minimum cost must equal the planned (i.e. design) life of the machine. Cost thus becomes the reliability apportionment criterion.

Cost is of course the implicit reason for the graded reliability in contemporary machines, but cannot be evaluated explicitly, because the proportion of failures (i.e. the reliability) is indeterminate using contemporary design practice. This is another feature that is subsumed into the factor of safety in contemporary design practice. Both the explicit and the implicit costing techniques do, however, result in different parts having different reliabilities at the design life.

It should also be appreciated that although reliability has been apportioned so that the cost is a minimum for the given layout or morphology, it does not follow that some other layout has not a lower cost. Others have to be evaluated using the usual trial-and-error technique to obtain the most acceptable design.

Cost parameters are so varied, and so many contributors are involved, that their evaluation cannot be treated on a general basis in this book, but must be a matter for consensus among those involved with the product. So too must the design life. It must not be so long that the machine becomes obsolete before wearing out, or so short that it wears out before a redesign is warranted (or the need for it ceases to exist in either case). These are all matters for judgement, and eventually ones that the design review body must get its teeth into – very firmly. Finally, one recalls from section 10.2 that if cost is the design criterion, then the reliability cannot be specified at the design life, nor indeed its variation over the machine's useful life.

So, what if the minimum cost does not result in the reliability specified for the machine? If the specified value has been based on experience of a similar machine operated in similar circumstances, the difference may not be too great, and some compromise should be possible. I should have thought, though, that the compromise must always be biased in favour of the minimum total cost. If one is concerned with an *ab initio* design of a complete machine, for which the specified and minimum cost reliabilities are far apart, as might well then be the case, a complete reappraisal of the specification and the design is clearly called for. The troublemaker here is the originator of specifications who thinks that it is a 'good thing' to call for ever higher reliabilities regardless of the economics. Even so, one would hope that in general the evolutionary approach to design described in the first section of this chapter was the operative one. In that case, specified and minimum cost reliabilities are likely to be acceptably close to one another.

Finally, in this section devoted to reliability apportionment, it may not be inappropriate to introduce a note of realism. Apportionment is based on cost, and as shown in the last section costing can be very imprecise. Some machines have a life expectancy of 25, perhaps 40, years or even more. Anyone who thinks that they can correlate the value of money over such a period is living in cloud-cuckoo-land. There are other constraints. In my experience nearly all machines are weight or volume limited, linear dimension limitations are also by no means unknown, and compatibility with existing hardware or procedures is a notorious constraint: the timescale for design may be circumscribed. Sales hype, even from within the designer's own organisation, can undermine the best design intentions. Then, although we all like to pretend that we are sophisticated and base our purchases on whole-life costs, in fact first cost is often a significant parameter affecting the final choice; of individuals, through small and large businesses, to governments. It's all a bit murky. Such aspects are, however, as important as the technial ones in defining the detail design of the parts of a machine. Design must embrace them all. It would not be surprising, therefore, if a little bit of the 'what we did last time' found its way into reliability apportionment: for example, short-term rather than a full-life economic optimum would be sought, or reliability would be based on some other non-economic criterion. If common sense and technical evaluation differ, it is time that a closer look was taken at the whole situation. This is not an easy task when the technical difficulties examined in the previous chapters are compounded with costing. But looking on the bright side, it is perhaps the breadth of its activities and the variety of circumstances that we encounter that make design so interesting.

10.4 The strategy

To sum up this chapter, it is asserted that any design strategy should encompass the following salient points, which have been highlighted in the preceding sections:

1. Load distributions become known only when the product (or a very similar one) is in the hands of the intended operator in actual use in the field. Reliability, i.e. failures, can only be properly assessed in these circumstances.
2. Consequently, target reliabilities not initially attained on introduction into service or subsequent reliability improvement must be achieved by economically correcting field failures as they arise. When this is done to *all* failures, a fully acceptable product so far as reliability is concerned will eventually be formulated.
3. Any organisation wishing to make reliability one of its selling points must be able to demonstrate continuous in-service reliability growth.
4. Design data for future products can be derived only from a full analysis of all incidents encountered in the field.
5. Prototype development or experimental design for reliability is both very costly and very time consuming, as witnessed by well-established values for the Duane index. While some prototype development is essential, major prototype development cannot economically replace good design backed up by continuous in-service development. Relying on development to bring a cursory design up to specification in a short time scale is an agenda for high cost or reliability shortfall.
6. It follows that major redesign should be avoided whenever possible, otherwise prototype development or subsequent continuous development will be overburdened; design for reliability is an evolutionary not a revolutionary process. Exceptions could arise when significant performance improvement is possible, or when a realistically calculated risk can be justified. A completely new customer demand obviously falls outside this requirement.
7. Maintenance costs are also best derived from field experience, but hidden costs should not be forgotten. They could well be more important.
8. Both machines and parts have an optimum reliability for a given product life, at which the initial cost and maintenance plus running costs are balanced so that the total or whole-life cost is a minimum.

9. A machine should obviously be designed for minimum total cost at its planned life, and in order that this itself should be minimised all constituent parts should have their minimum cost at that life. This forms a part's reliability apportionment criterion. It is a necessary but not always sufficient criterion: a totally different morphology could be more economic.

Having identified a method by which parts can be designed for a prescribed reliability – or minimum cost – at a specified life (or for an infinite life in the case of stress-rupture mechanisms), and having shown how each part can be allocated a reliability in line with the machine's specification, we have reached the end of our search for a design methodology that can meet the demand for a numerically specified reliability and the extent to which that demand can be met. The methodology is not far from current practice, and therefore should be acceptable to contemporary designers. The objective has been achieved, but a lot of assumptions have had to be made on the way. The next and final chapter is therefore an overall review of the resulting situation.

References

Carter, A.D.S. (1986) *Mechanical Reliability*, 2nd edn, Macmillan Press, Basingstoke.

Duane, J.T. (1964) Learning curve approach to reliability monitoring. *IEEE Transactions on Aerospace*, **2**, 563–566.

Consumers Association (1994) *Which? Guide to New and Used Cars*, Consumers Association, London.

Hastings, N.A.J. (1967) A stochastic model for Weibull failures. *Proceedings of the Institution of Mechanical Engineers*, **182** Part 1 (34), 715–723.

Vinal, P. (1991) Oceanic twinjet power, Royal Aeronautical Society, London.

Appendix 1: Water pump reliability specification

The reason why the specification cannot be met is the trough in the mission reliability between about 10 000 and 15 000 miles. The trough arises from the peak in the hazard curve: see Figure 6.36(c) for an example. If the stress is reduced in an attempt to improve the reliability, the failure-free life will be increased, but the trough remains. The lower limit of the trough may be slightly raised, or perhaps not noticeable, depending on the location of the design on the s–N curve. Substantial reduction of the stress may be necessary to move the trough beyond the full life, or to increase its lowest value to the specification. It will incur a significant increase in weight and volume, and in cost. This is generally not acceptable. The manufacturer then has several options open to him:

1. He can ignore the trough and offer the existing pump as meeting the specification – slightly dishonestly. He might remember that it would be a sophisticated operator who could detect the trough, and it would take some
   time to accumulate the data!
2. If the customer requires test evidence that the design can meet the specification, it would become necessary to 'adjust' ('cook' in everyday lan-
   guage) the field data to eliminate, or at least reduce the trough; e.g. by
   fitting different distributions to the data, etc.
3. An alternative would be to increase any warranty period so that it includes
   the trough, though the manufacturer would have to increase the price of
   the pump if he wished to make the same profit.
4. An air of respectability can be given to option 1 by quoting the average or
   mean breakdown rate over the planned life. It is the technique adopted in
   the epiphenomenological approach, which is virtually always based on a
   constant 'failure rate'.
5. It is not unknown for organisations to refuse to quote against specifications that contain a reliability requirement.

This is of course a specification, not a design, problem. One can always
design to meet a specification, but whether the written specification represents the intention of the customer is another matter. One would hope that
any misunderstanding would be resolved at the first design review, though it
is not always so in practice.

There is a fundamental aspect of this situation that should always be recognised. The shape of the mission reliability curve in Figure 10.5 is a direct
consequence of the shape of the $s–N$ curve. Conversely, it is the shape of the
$s–N$ curve, or in most practical cases its limit shape, that is being defined by
any mission reliability specification, or indeed any specification that defines
the reliability at various stages of the product's life. This limit shape may
conform to actual $s–N$ curves, or most likely it will not. In the latter case the
designer is being given an impossible task. The writers of specifications very
often do not realise the consequences of their actions! In my opinion the
reliability (or cumulative failures) should be specified only at one age, which
will nearly always be the planned life, unless the originator is fully conversant with mechanical reliability and the design process. This would allow the
designer the freedom to chose the most appropriate $s–N$ curve, i.e. material,
to optimise the design for other desirable features.

Appendix 2: The reliability of a series system

The subject matter of this appendix is really more a part of statistical reliability theory than design, but it has been included because of its relevance to
reliability apportionment.

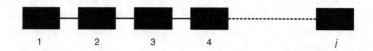

Figure 10.6   The reliability block diagram (RBD) for a simple series system.

Nearly all reliability texts state that the reliability of the series system shown in Figure 10.6 will be given by

$$R = R_1 \times R_2 \times R_3 \times \ldots \times R_j \qquad (10.12)$$

where R is the reliability of the system; R_1, R_2, R_3 etc. are the reliabilities of the blocks making up the system; and j is the number of blocks. Figure 10.6 may be considered as representing a system with each of the blocks representing a subsystem, or as representing a machine with each of the blocks representing a part of the machine. The only difference in these interpretations is that there are likely to be very many more parts in a machine than subsystems in a system.

Equation (10.12) is derived from the statistical product rule to meet the condition that for the system/machine to be functional all the blocks must also be functional. But the product rule is valid only if all the blocks are truly independent. In the case of a machine this condition may not always be satisfied; indeed it rarely is. The load applied to the machine is common to all the blocks (or the failure-inducing load on each block, i.e. part, is a function of this load), the age of the blocks is common or related through any maintenance procedure, and the life distributions are – or should be – related by cost.

The author has shown mathematically – see Carter (1986) – that if the part is subject to a stress-rupture failure mechanism, then equation (10.12) is valid only if the loading roughness is zero: i.e. the machine is subject to ideally smooth loading. If it is subject to infinitely rough loading (loading roughness $= 1$), then the reliability of the machine is equal to the reliability of the least reliable part. Intermediate loading roughnesses give rise to intermediate results.

For wear processes (not considered in the above treatment), the failure of individual parts will be represented by the probability density functions $f_i(N)$, where N is the number of load applications to failure. Let us first consider the reliability of a new machine operating in an infinitely rough loading environment; i.e. loading roughness $= 1$. Then from equation (4.9), either the standard deviation of the strength and hence the damage resistance is zero, or the standard deviation of the load is infinite. Both these conditions imply that

$$f_i(N) = \infty \qquad \text{for } N = (N_F)_i$$

$$f_i(N) = 0 \qquad \text{for } N <> (N_F)_i \qquad (10.13)$$

for all i, as illustrated in Figure 10.7(a). In drawing this figure it was assumed that $(N_F)_i$ increases with i. No generality is lost by this assumption,

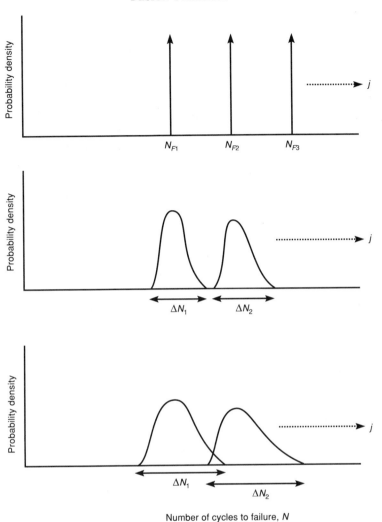

Number of cycles to failure, *N*

Figure 10.7 Probability density functions to failure of a machine's parts: (a) loading roughness = 1; (b) loading roughness < 1 (no interference); (c) loading roughness < 1 (interference).

though it may require some renumbering of the blocks in Figure 10.6. It may be argued that N_F should be the same for all parts (shades of Wendell Holmes!), but this is a counsel of perfection that is unlikely to be achieved in any practical design. Because the number of loads must increase monotonically in steps of unity from zero, it is evident from Figure 10.7(a) that the least reliable part must be the first and only part to fail. The second assertion follows because, for $N > (N_F)_1$, 100 per cent of the least reliable part, and hence 100 per cent of the machines, must have failed. No $N > (N_F)_1$ can be reached. The reliability, measured as $(N_F)_i$ for $i = 2, 3, \ldots j$ of all the other parts, is irrelevant, as indeed is the number of parts: hence the reliability of the machine is equal to the reliability of the least reliable part. This is the same conclusion that was reached for stress-rupture in these circumstances.

If the loading roughness is less than unity, the failures of each part will be spread over a finite range of N, say δN. An example is shown in Figure 10.7(b). It is clear for this particular location of the pdfs that as n is increased the reliability of the machine represented by Figure 10.7(b) will be equal to the reliability of the least reliable part. The least reliable part is here defined as the part for which $N_F = N_\delta$ is the smallest, where δ is a small percentage failure approaching zero. As drawn, 100 per cent of the least reliable part, and hence of the machines, will have failed before any other part could fail. So once again the reliability of the other parts and their number is irrelevant.

If, however, there is 'interference' between the failure distributions of the least and the next least reliable parts, or perhaps between the least reliable and several other parts, then the possibility of failure of any of the parts at risk has to be taken into account. Figure 10.7(c) illustrates the former case, for which quantitatively

$$R(t) = R_1(t) \times R_2(t) \tag{10.14}$$

or in general

$$R(t) = R_1(t) \times R_2(t) \times \ldots R_i(t) \tag{10.15}$$

where i is the number of interfering parts. The non-interfering parts' reliability and their number are again irrelevant.

The number of interfering parts will be a function of both the loading roughness and the apportionment of reliability among the parts. It is not until all the parts are interfering that equation (10.12) becomes applicable.

Unfortunately, it is not clear from equations (10.14) and (10.15) that, unless the interference between the failure distributions is substantial, the reliability of the machine for practically relevant values of the machine reliability is still equal to the reliability of the least reliable part. For example, Figure 10.7(c) indicates that about 95 per cent of the least reliable part, and hence the machines, will have failed before the next least reliable part can make any contribution to the failure process. One is not likely to be much interested in the behaviour of the remaining 5 per cent. Even with substantial interference it would be remarkable if the N_δ for many parts were the same, or indeed if the N_δ value for the next least reliable part were less than the N_5 or N_{10} values for the least reliable part. Thus in the majority of machines the machine reliability of practical interest will still often be equal to that of the least reliable part.

The above description refers to a non-maintained machine, or to a new machine up to first failure. Most machines will, of course, be repaired when a failure occurs, and the pdfs in Figure 10.7 have to be modified to correspond to the original/repaired age mix at any time (number of load applications). The mix will depend on the maintenance policy as well as on the machine design. Evaluation can become tedious if carried out by hand, but computer evaluation is relatively easy. For example, the Hastings program – there are many others – will determine the age mix for a part once the maintenance policy is known. The basic theory underlying the Hastings program has been given by Hastings (1967) and by Carter (1986).

It will be appreciated that the limited number of interfering parts constitute Pareto's 'vital few'. Although with continued maintenance it is obvious that ultimately the reliability of all parts will have to be considered, Pareto distributions imply that even with maintenance only a small number control the reliability. This field evidence cannot be ignored. It does seem to me that for design purposes, and within the limits of accuracy that we can expect at the design stage, it will be sufficient to take into account only those parts whose pdfs are expected to interfere (Pareto's vital few) when evaluating the reliability of a machine from its constituent parts.

11
A final review

This book has been concerned with the achievement of a quantifiable (i.e. statistically defined) mechanical reliability at the design stage. However, no attempt has been made to cover the epiphenomenological statistical aspects of design applied to systems, which at present is possibly the most widely practised form of statistical design. There are two reasons for this. First, the basic principles are well established, and have been covered adequately in a number of good books. Hunsley (1994) provides a good introduction to practical mechanical systems reliability for those unfamiliar with this activity, while O'Connor (1991) gives a good general introduction to reliability practices in the electronic and electrical as well as the mechanical fields. There are many others, as well as more specialist and 'advanced' works. There is no need for yet another repetition. Second, the traditional methods of estimating reliability are coming under closer and closer scrutiny, even in the electronic industry for which they were first devised, as the limitations of the epiphenomenological approach become more apparent. It is not at all clear at present what role they will have to play in the future.

If such techniques, which are really devised to side-step the real mechanical engineering activities, are firmly rejected, it is surprising how quickly one reaches the heart of mechanical reliability: detail design. This was recognised in principle at least 300 years ago – 'for want of a nail . . .' and all that. It is the inevitable conclusion of a close appreciation of mechnical design, and so this book developed into a study of the achievement of mechanical reliability by detail design. I have no regrets about that. Even the most complex system fails only because a part has failed.

As stated in the opening chapter, mechanical reliability depends only on the shape, size and material of construction of every part. Tremendous advances have been made in materials science and stress analysis, especially when making use of modern computers, with obvious relevance and benefits to these three vital characteristics. This is all well documented. But to use these sciences via old-style empirical techniques is certainly 'putting new wine into old bottles'. So this book has become a search for 'new bottles'.

The guiding principle in compiling the book has been to establish as accurate a physical model of the real situation as possible, and to use this as the basis for any new design methodology. Thus reliability is a consequence of the variation in the behaviour of the individual items comprising a product. Variable behaviour arises directly from the variation in the strength or damage resistance attributes of the numerous bits and pieces from which the product is made, and indirectly from the variation in the load or environment either self-imposed or imposed by the user. Ultimately, there really can be no doubt that such variation is best evaluated by statistical parameters that have been specifically devised for that purpose, and which have been tried and tested in so many different applications over a long period of time. It was shown in Chapter 3 that there are some special problems in doing this for the quantities involved in mechanical reliability, and these problems should never be underrated. It is not my intention to do so now. Every designer who uses statistics must be fully aware of the pitfalls. The problems are associated with the lack of extensive data and the necessary consequent extrapolation rather than with any fundamental issue, though the latter is open to misinterpretation in such circumstances. In some form they confront any mechanical design methodology. There is no reason why statistically oriented design methods should be particularly debited with this problem though equally they cannot be relieved of it. Acknowledging this difficulty the subsequent chapters of this book showed how statistically defined loads, strengths and damage resistances could be incorporated into an acceptable methodology. In so doing, conventional epiphenomenological statistical methods were deliberately discarded. They have no physical meaning. Instead, the methodology used statistical representation, but only as input data. Those data are then manipulated in accordance with the physical behaviour of the material and the interrelationship between the physical properties derived from materials science to put them in a form that can be used in a more or less conventional design process.

Rereading the manuscript of the book in that context I feel that the original objective has in a large measure been achieved, but not with the certainty that one would have liked in all areas. So far as possible, a true scientific philosophy has been adopted. At each step, support for any hypothesis by test evidence has been sought before proceeding to the next. Unfortunately, because of lack of data, critical supporting evidence is sometimes missing, and when available, corroboration in one case does not mean universal corroboration. Even so, the conclusion that strain-hardening is near deterministic is considered a sound one. The theoretical argument has a firm basis, and is backed by test evidence from the best set of stochastic material data that is available. Likewise, I believe that the model of constant strength throughout substantially the whole of an item's life, followed by sudden death, is fully valid. It is the picture that would be held by most engineers for deterministic behaviour, and there is a great deal of evidence to support that view. It is difficult to see how the behaviour could be selective. There is thus strong evidence for believing that it is equally valid for every item of a distributed population, and is backed by test data, albeit on only one material. There is no reason to believe that it is an exceptional result, though a very big query must be set against composites. As these are likely to become more and more important engineering materials in the future, there is clearly a large gap in our knowledge. For the general run of engineering design, however, sudden death would seem to be a fully acceptable model. The conclusion that the standard deviation of the fatigue damage resistance is constant throughout the whole life of a part (i.e. for all values of N) takes us on to much less firm ground.

Although based on modern accepted (but deterministic) modelling of fatigue by crack propagation, the extension to the behaviour of a distributed population is, in my considered opinion, just a plausible argument that could be advanced in support of established test data – were that at hand. It cannot be regarded as an *ab initio* reasoning that has stand-alone authority. Too many approximations have had to be made that may, or equally may not, be true. Worse still, there are hardly any comprehensive test data that can be used to support the hypothesis. The evidence available is all very 'bitty'. The difficulty is that a very large coordinated programme indeed would be necessary to provide the data on even one material. That material would have to be fabricated under strict, but not artificial, quality control into a number of real representative test components, and the test programme would have to be statistically planned to provide results at a high confidence level, and systematically executed. We shall have to wait a long time before we see such a test, let alone the large number required to generalise. One doubts if Bore's work (quoted earlier) falls into the above category. In fact, I am unable to decide whether Bore's firm assertion that the standard deviation ('scatter' in his terms) is constant for all N is the result of considerable insight, an inspired guess, or is indeed based on comprehensive test evidence. On the other hand, the evidence based on Cicci's work is, I think, very sound so far as it goes, but it does not go far enough. Kecicioglu's work is statistically inconclusive, although well orientated. The most telling support is provided indirectly by the agreement between the 'safe' $s–N$ curve for aircraft structural material and the calculated curve. This implies that the standard deviation is constant. If the calculated curve is in error, so too is the empirical curve. I cannot think that the latter can be far in error from such a reputable source.

Neither must it be forgotten that the critical assumption underlying the theory (the deterministic nature of the propagation equation parameters) also underlies the application of the Paris equation, which is extensively used, and gives good agreement with experience. It may be considered that the assumption of a deterministic d (distance to the strongest barrier) lies outside this analogy, but is the soundest assumption. In any homogeneous well-finished item there will be a large number of incipient cracks, each with its own d. But because this number is large, the distribution of d in any item will be close to the population distribution, and therefore the same for all items. It follows that the effective value of d must be virtually the same for all items: i.e. d is deterministic. As a general principle, if any material property is a function of a large number of some randomly dispersed microstructural feature, so that the standard error of the mean and the standard error of the standard deviation of the sample of the feature's contribution to an item's property is small, then that property will be near-deterministic.

The application of the theory to other failure mechanisms (as in section 6.3) is on even less firm ground than that for fatigue. Indeed, that section is little more than speculation. It may, however, be worth recording that design based on statistically defined data described in the previous chapters does not depend on the constancy of the standard deviation, either for fatigue or for any other failure mechanism. The required $s-N_F$ curve can always be found by test. The practical value of the theory is that it allows design data to be obtained from limited test data, and permits the latter to be checked for consistency.

Rereading the manuscript, I am inclined to think that perhaps too much attention has been paid to fatigue, though fatigue specialists may observe that there is a lot more to the subject than has been covered in this book. For example, such features as corrosion fatigue, surface treatment, temperature cycling, the effect of material grain size etc. have been completely ignored. Even a deterministic representation of these topics presents problems. Just so; but it has been shown that if the current median $s-N$ curve embodying these aspects is available, together with the distribution of the corresponding ultimate tensile strength, all further data necessary for statistically specified design can be calculated. Very little that is new is required. One must acknowledge, though, that fatigue is generally considered to be the outstanding source of failures. Some say that as much as 90 per cent of failures are due to fatigue, though my own analysis of field data (limited to military equipment, but otherwise extensive) suggests that erosive wear and corrosion are equal culprits. I would agree, however, that fatigue is seen as the bogeyman. For this reason research has concentrated heavily on fatigue, and more data – at least superficial data – are available on fatigue than on any other failure mechanism. It was inevitable that the treatment in this book should follow the same pattern.

My reflections on Miner are more sanguine; there's nothing else. The expected adverse error appears to be only 20 per cent (see equation (7.15)). What's that in design!

A fair conclusion from this review is that the physical behaviour upon which the statistical methodology has been based does have some experimental support, though a scientist or statistician would assert that the verdict must be 'not proven'. It is as an engineer rather than a scientist that I claim that the test evidence, *taken as a whole*, justifies the description of material behaviour described in this book, and in turn justifies the design methodology that has been developed. In particular the evidence, although 'bitty', does interlock: see for example the case study of Chapter 7. I do not doubt that there are secondary perturbations that have not been explored, but it does seem pointless to pursue these smaller effects when much more significant aspects have still to be resolved. I have particularly in mind duty cycles or load distributions and the cost data on which optimisation ultimately depends. At least the statistically based methodology forces a consideration and some evaluation of these aspects during the design process. Current empirical methods just sweep them under the carpet.

Having been so equivocal regarding the alternative implementation of worst-case stress-rupture design by factors of safety or statistical representations, it was with some relief that I at last came down on one side: the latter, as it turned out. In retrospect, I do not think that there is any alternative, if design is to advance as a coherent concept. One must regret that the statistical approach cannot deal with the demand for very high reliabilities. It does, however, demonstrate why this is not practically possible, and the circumstances in which it arises. In practice, high operational reliability can always be obtained by on-condition maintenance or fail-safe techniques: see equation (8.1). The demand for very high reliability for every part, particularly when exploited by emotional safety aspects, is thus a red herring so far as choice of methodology is concerned. No methodology can predict the adventitious circumstance that has not been previously experienced and of which there is no evidence. When bound by the restrictions of time and cost, evaluating at some 5 or 6 standard deviations from the mean some material property or some load can only at best be intelligent speculation. The factor of safety has no magical powers in this respect. It seems to me that the statistical approach is realistic, and hence to be preferred to a 'head-in-sand' approach based on factors.

But prejudices die hard! I can see that many older designers would wish to stick to the empirical factors with which they have so much experience. It is very understandable. I can see that as one tries to represent each of the inputs statistically, the imponderables become greater and greater. I have no doubt that all material properties could be evaluated statistically. I have no doubt that s–N_F curves for any value of F could be calculated for all the common wear failure mechanisms, should we wish to do so. But doubts arise when representing load distributions. There are a significant number of unusual operators who cannot be entirely ignored. Doubts grow at the next stage, when eccentric maintenance policies have to be accommodated. Finally costing, on which everything must ultimately be based, involves goodwill and other intangible aspects besides a multitude of standard everyday practices giving rise to different cost parameters. As the going gets harder one can see the attraction of bundling all the imponderables up into one empirical factor. At least there is only one thing to think about then. As a person who was brought up on factors of safety and used them all his professional life, their simplicity appeals to me. However, if we are to make any progress the bundle has to be unbundled, and each of the constituents correctly modelled. Is every unusual operator to be catered for? Can we optimise costs for every maintenance policy and emotion outburst? I do not think so. Many must be included of course: others deliberately and openly excluded. Refined machines have to be designed for a specific duty – 'horses for courses' as the general saying has it. This outlook is embodied automatically into the design by such modern techniques as QFD (quality function deployment), so there is no need to pursue them here, except perhaps to note that the effective implementation of these techniques requires the unbundling anyway: see for example Cross (1994) and further references contained therein. It should also not be forgotten that the operator and maintainer have a duty to to ensure that these activities conform to specification. They carry as much responsibility for the ultimate reliability as the designer, and cannot opt out.

The final issue, so far as I am concerned, is the transparency of the design methodology based on statistically defined input data. Every feature of the design has to be separately and quantitatively defined. All the data can, ideally, be checked by test: even if it can prove prohibitive in time and cost actually so to do. Of more importance, all those input data can be and should be agreed by all involved, and can be subsequently audited; and, given those input data, the outcome is independent of any subjective manipulation. Initial clarity is essential. Having listened to many designers explain to me what their factor of safety is supposed to do, often trying to justify the unjustifiable, I am convinced that clarity is the one thing most contemporary design techniques based on empirical factors do not possess. I thus come down firmly on the side of a design methodology based on statistically defined input data; I hope that the reader agrees. The factor of safety has served us well for over a century, and among others I should be sorry to lose a familiar tool, but it is fundamentally flawed and must go – eventually.

For no one would – or should – dream of changing from one method to another overnight. The old and the new methodologies must be used together until a permanent change is, or is not, justified. We are fortunate in that the two methodologies are not all that dissimilar. Using a particular $s-N_F$ curve in place of an empirical 'safe' $s-N$ curve is not very different, so far as implementation is concerned, though quite different in underlying philosophy, derivation, interpretation and, more particularly, in potential. Parallel implementation should present no problem.

One possible illusion must be firmly renounced. Using statistically based design techniques does not solve the whole problem. Design has endless ramifications outside reliability, and is essentially an iterative process with unknown interactions. They will always exist, and thus never remove the necessity for prototype testing. Every design must always be proved by test. This always has been true and always will be. What we may reasonably expect, however, is that statistically based design will be much closer to the desired end-result: quantified reliability/minimum cost. It follows that though we cannot expect to eliminate prototype testing, we can reasonably expect the amount of such testing, that very expensive activity if reliability is to be verified, will be considerably reduced.

Finally, the reader is reminded that the subject matter of this book forms but a very small part of the total design activity. But in bringing his or her original concepts to fruition, the designer must conform to the appropriate technological discipline – the grammar and syntax of the trade, as it were. It is often the overriding demand. I have seen many a promising innovative design come to grief when each of the parts has been sized for adequate strength. It is to that aspect that this book has been addressed: particularly to the means by which a more, and ever-more, demanding requirement can be met. This may seem a bit grandiose, especially when 90 per cent of design is mundane copying. But even standards and codes must be formulated, and must acknowledge that the customer is demanding a quantified reliability so that he can optimise his own activities. The implementation of the methodology is in fact a lot easier when formulating codes and standards than in original design. Codes and standards cover such a wide and heterogeneous set of circumstances that precise distributions serve no purpose. The normal distribution is usually good enough. So most of the problems associated with defining the extremes of distributions disappear. The errors will, or could be, very large, of course: but that is inherent in the nature of codes and standards. If the codes and standards are to be acceptable, the errors, too, must be acceptable: i.e. take the form of overdesign. But such overdesign should be rational, not what some forceful character/organisation thinks it should be. I cannot say that this kind of design inspires me greatly, but it may be adequate for much run-of-the-mill stuff. For more precise, enlightened and optimised design, the designer will often be on his own, and have to establish and verify his own data, though I think he should demand as of right a more accurate specification of the design in all its ramifications than is now customary – or else the resources to find them for himself. I believe this to be well worth the effort, for there still seem to be rich rewards in optimising reliability to achieve the top performance requirements cost-effectively. It has a down-to-earth, as well as an aesthetically satisfying, objective. Good design always has been, and I think always will be, the key to success, both commercial and technical. It should be pursued vigorously in all its aspects. This book has looked at just one and, I hope, has reached some useful conclusions.

References

Cross, N. (1994) *Engineering Design Methods*, 2nd edn, John Wiley & Sons, Chichester.

Hunsley, C. (ed.) (1994) *The Reliability of Mechanical Systems*, 2nd edn, Mechanical Engineering Publications, London.

O'Connor, P.D.T. (1991) *Practical Reliability Engineering*, 3rd edn, John Wiley & Sons, Chichester.

Index